新编职业英语系列

新编
计算机英语

中等职业学校职业英语教材编写组

高等教育出版社·北京

图书在版编目（CIP）数据

新编计算机英语 / 中等职业学校职业英语教材编写组编． -- 北京：高等教育出版社，2020.7（2023.11重印）
ISBN 978-7-04-054346-9

Ⅰ.①新… Ⅱ.①中… Ⅲ.①电子计算机－英语－中等专业学校－教材 Ⅳ.①TP3

中国版本图书馆CIP数据核字(2020)第111350号

| 策划编辑 | 袁艺杰 | 责任编辑 | 康冬婷　王宇茜 | 封面设计 | 王　洋 | 版式设计 | 孙　伟 |
| 责任校对 | 李　森 | 责任印制 | 朱　琦 | | | | |

出版发行	高等教育出版社	网　　址	http://www.hep.edu.cn
社　　址	北京市西城区德外大街4号		http://www.hep.com.cn
邮政编码	100120	网上订购	http://www.hepmall.com.cn
印　　刷	天津鑫丰华印务有限公司		http://www.hepmall.com
开　　本	787mm×1092mm 1/16		http://www.hepmall.cn
印　　张	16.25		
字　　数	388千字	版　　次	2020年7月第1版
购书热线	010-58581118	印　　次	2023年11月第5次印刷
咨询电话	400-810-0598	定　　价	35.00元

本书如有缺页、倒页、脱页等质量问题，请到所购图书销售部门联系调换
版权所有　侵权必究
物　料　号　54346-00

致 同 学

  亲爱的同学们，祝贺大家即将迎来新学期的学习生活。我们根据教育部于2020年3月颁布的《中等职业学校英语课程标准》中"职业模块"的要求，为大家编写了这本《新编计算机英语》。希望大家在《英语》（基础模块）学习的基础上，进一步学习英语语言知识和计算机相关文化知识，提高在相关行业中运用英语进行交流的能力，并在学习中感知中外思维间的差异，进一步提高计算机素养和人文素养。我们也希望大家能在学习中养成良好的学习习惯，发展自主学习能力，成为具有正确价值观和职业观的高素质技术技能人才。

  本册教材以计算机相关行业主题为主线，以交际任务为驱动，基于中职英语学科核心素养进行编写，具有以下特色：

  • 将语言知识和专业技能有机地融为一体。本册教材采用多技能大纲的模式，以专业技能为主线来组织教学内容，融听、说、读、写四项技能训练为一体，使同学们能在未来的职业场景中运用英语完成基本的沟通与交际任务。

  • 各单元主题职业特色鲜明，以模块化的结构展现学习内容。本册教材共14个单元，每个单元的主题都与计算机相关行业所需要具备的专业知识和关键能力相关。内容安排由浅入深、循序渐进，由计算机的组成、计算机的软硬件配置、常用计算机应用软件、计算机网络等主题构成。各个单元的内容既彼此相关，又相对独立，同学们可根据自己的学习程度与学习需求，完成相应模块的学习。

  • 选材注重知识性与实用性、可思性与趣味性、经典性与实效性的结合。本册教材立足于本土化的职业场景，力求将语言学习置于真实的职业场景之中。所选语言材料难度适中，语言精练生动，具有时代特色；配图精美，版式活泼。

  • 以多样化的活动形式和任务驱动的方式，借助情境化的场景培养大家解决问题的能力，让同学们在学习中体验，在体验中发现，在发现中学习。

  本册教材每个单元的编写体例结构如下：
Unit Goals单元目标
  基于目标导向，每个单元首先明确在语言技能方面所应达到的要求，以使同学们预知在学完每一个单元后能够用语言去做什么，希望大家在课程的学习中能够有意识、有目的地进行学习。
Lead-in导入
  导入部分以图文并茂的形式，列出了一些与本单元主题相关的话题，本部分的作用是为大家在新旧知识之间建立关联，为后面的学习和训练做好准备。
Listening and Speaking听说
  听说部分围绕单元主题，采用典型的、简练的对话，通过交际情境展示内容，同时为会话表达和口语练习提供范例。听力练习后是互动性的角色扮演练习，帮助大家掌握并灵活运用对话部分所学的内容。

Reading 阅读

阅读部分包括两篇与单元主题紧密相关的课文，以专业知识为话题内容，文字浅显易懂，配有简明的文前、文后练习及注释，生词表除基本词汇外，还包括拓展词汇。阅读部分有效地融合了语言知识学习和专业知识学习，旨在培养大家在未来的职场中能够理解常见的计算机相关行业阅读材料的能力。注释部分的主要术语采用中英双语对照的形式，进一步提升大家在计算机相关行业职场中使用英语进行交流的能力。

Do You Know? 知识拓展

知识拓展部分旨在帮助大家拓展计算机相关行业的相关知识。

Language Practice 语言练习

这部分的所有练习与单元主题相关，通过设计多样化的语言活动，帮助大家运用所学的语言知识，提高在计算机相关行业中的语言交流能力。

Writing 写作

写作部分是与课文内容相关的一篇简要的短文写作，旨在培养大家运用所学知识进行初步的专业内容写作的能力。

Grammar 语法

语法部分以图表形式简洁直观地呈现新的语法知识并配有相关练习，重点突出，便于掌握。

Fun Time 快乐时光

在经过了一个单元的学习后，我们为同学们选取了与单元主题相关的一则小笑话，让大家在学习之余轻松一下，感受语言学习的乐趣。

Project 项目活动

通过难度逐步升级的三个步骤的小组活动，项目活动部分引导大家独立或与同学合作完成课后的调查、实践或学习任务，旨在激发大家有效运用所学语言知识和专业知识进行讨论、分析和解决真实职业场景中问题的能力。

Self-checklist 自我评估表

自我评估表帮助大家对学习效果进行自我检查，并构建个人学习的成就档案。

此外，扫描本书郑重声明页上的二维码或访问二维码下方的链接，可以查看并下载本书配套的录音及教学资源。

本教材由暨南大学外国语学院赵雯教授担任总主编，东北大学外国语学院姜蕾、刘春阳任主编，参加编写的有东北大学郝丽霞、张宜波、武芳芳、王祁和王勃然。暨南大学黄卫祖教授审阅了全书。在编写过程中，我们也听取了多所职业院校专业英语教师的意见，在此一并致谢。

由于作者的知识与水平有限，不当之处在所难免，恳请大家批评指正。衷心希望本书能成为同学们的良师益友。

编者
2020年5月

Contents

Unit 1	What Is a Computer?	1
Unit 2	Hardware	16
Unit 3	Software	32
Unit 4	Basic Computer Skills	48
Unit 5	Word Processing Software	65
Unit 6	Spreadsheets	82
Unit 7	Database	100
Unit 8	PowerPoint	118
Unit 9	Desktop Publishing Software	135
Unit 10	Computer Use	151
Unit 11	Network	167
Unit 12	The Internet	184
Unit 13	Some Important Issues	199
Unit 14	Future Computers	216
Words & Expressions		233

Unit 1

What Is a Computer?

Unit Goals

In this unit, you will be able to
- understand listening materials about what a computer is;
- talk about computers;
- understand articles about basic computer knowledge;
- write a short paragraph about computers;
- demonstrate your knowledge about what a computer is.

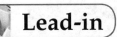

Picture matching.

1. mouse   2. keyboard   3. monitor   4. CPU   5. router

Listening and Speaking

Listen and complete.

a. What's a computer?
b. Sure.
c. It's really fun.
d. Oh, I think it's everything.

What Is a Computer?

Different people have different ideas about what a computer is. Sarah, a student from a computer school, is asking people the question.

She is talking with her friend, Tom.

① Sarah: Hi, Tom. What's a computer?
    Tom: Well, it's just a game machine. _____

She is asking her English teacher, Ms Lee.

② Sarah: Good afternoon, Ms Lee. _____
    Ms Lee: _____ It's a great help to me.

She is asking Miss Wang, a secretary in an office.

③ Sarah: Excuse me, Miss. Could you tell me what a computer is?
    Miss Wang: _____ It's a useful tool for me. But it always keeps me busy. I don't like it at all.

Role Play

A: Excuse me, Miss / Mrs / Mr ... What's a computer?
B: Well, it's a game machine / useful tool ...

Unit 1 What Is a Computer?

Words & Expressions

mouse /maʊs/ n. 鼠标
keyboard /ˈkiːbɔːd/ n. 键盘
monitor /ˈmɒnɪtə(r)/ n. 显示器
CPU (Central Processing Unit) n. 中央处理器
router /ˈruːtə(r)/ n. 路由器
secretary /ˈsekrətri/ n. 秘书

It's really fun. 它真的很有趣。
talk with 与……交谈
It's a great help to me. 它对我的帮助太大了。
not ... at all 一点也不

Reading

Text A

Pre-reading activities.

1. Picture matching.
   1) a supercomputer   2) a desktop computer
   3) a tablet   4) a laptop computer

a

b

c

d

2. Do you have a computer? If yes, tell your peers about it.

3. Find the following words in the text.

supercomputer   desktop computer   tablet   laptop computer

3

Reading Strategy

扫读 Scanning 也称寻读，是一种在阅读过程中查找人名、地名、年代、数据时经常会用到的阅读策略。通常阅读者在阅读时快速扫读全文，把与问题无关的词、句、段略去不读或不细读，只找到某个特定的词、词组或某个问题的相关答案即可。

Types of Computers

Computers can be found in many shapes and sizes and they are used to do many things. In this text, we will show you some types of computers.

Supercomputers are very big. They are usually larger than a classroom, and they are also very powerful.

Desktop computers are very popular and widely used. As the name shows they have most of their parts on the desk. Different parts are connected with either cables or Wi-Fi. Most families can afford one nowadays.

Notebook computers are also called laptop computers. They are small enough to be put on your lap or in your briefcase. People may carry them around, but they are usually more expensive than desktop computers.

Tablet personal computers are usually shortened as tablets. They have neither mouses nor keyboards. People touch or slide on the screen with their fingers to operate tablets and use pop-up keyboard for typing. People can also use stylus or digit pen to operate or input. Built-in handwriting recognition and voice recognition make input more convenient than before. Tablets are typically larger than smartphones. They are very popular among young people.

There are also other types of **mini computers** and you can find them around you, such as those in microwave ovens, washing machines and refrigerators.

Unit 1 What Is a Computer?

Words & Expressions

supercomputer /ˈsuːpəkəmpjuːtə(r)/ *n.* 超级计算机
shape /ʃeɪp/ *n.* 形状
powerful /ˈpaʊəfl/ *adj.* 强大的
connect /kəˈnekt/ *v.* 连接
cable /ˈkeɪbl/ *n.* 电缆
Wi-Fi /ˈwaɪ faɪ/ *n.* 无线局域网；无线网络
afford /əˈfɔːd/ *v.* 买得起；承担
tablet /ˈtæblət/ *n.* 平板电脑
screen /skriːn/ *n.* 屏幕
slide /slaɪd/ *v.* 滑动
stylus /ˈstaɪləs/ *n.* 触控笔
build-in /ˌbɪlt ˈɪn/ *adj.* 内置的

convenient /kənˈviːniənt/ *adj.* 方便的；省事的
pop-up /ˈpɒp ʌp/ *adj.* （计算机窗口）弹出的
smartphone /ˈsmɑːtfəʊn/ *n.* 智能手机

desktop computer 台式机
laptop computer 笔记本电脑
carry around 四处携带
handwriting recognition 手写识别
voice recognition 语音识别
digital pen 数字笔
such as 例如，像

True or false.
1. Supercomputers can be carried around.
2. Desktop computers are more expensive than laptop computers.
3. Different parts of notebook computers are connected with cables.
4. Voice recognition makes tablets convenient to use.
5. There are some mini-computers in microwave ovens, washing machines and refrigerators.

Text B

Pre-reading questions.
1. Do you know which part of a computer is the most important?
2. What do the letters of CPU stand for?

5

What Is a CPU?

If you want to buy a new computer, it is necessary to understand what a CPU is. The letters of CPU stand for Central Processing Unit, which is the "brain" of your computer. Without the CPU, you would not be able to play games, type research papers, or surf the Internet.

Sometimes people are wondering where the CPU is. In fact, you cannot see a CPU from the outside. You have to see inside to get a good look.

The first CPUs were used in the early 1960s. They were made as part of a larger computer. Once engineers figured out how to mass produce the CPU, personal computers became less expensive.

Since the CPU is one of the most important parts of a computer, no wonder that it is also the most expensive. If your computer is more than three years old, you may want to upgrade it to a new one. According to Moore's Law, the processor speed or overall processing power will double every two years. A newer, faster CPU will hence often enable you to be a better user.

Words & Expressions

central /ˈsentrəl/ adj. 中央的
process /ˈprəʊses/ v. 处理
unit /ˈjuːnɪt/ n. 单元
surf /sɜːf/ v. （互联网）冲浪，浏览
wonder /ˈwʌndə(r)/ v. 想知道；感到诧异
personal /ˈpɜːsənl/ adj. 个人的
upgrade /ˌʌpˈɡreɪd/ v. 升级

stand for 代表
in fact 事实上
figure out 想出；计算出
mass produce 批量生产
no wonder 毫不奇怪

Short-answer questions.

Answer the following questions according to Text B.

1. What do the letters of CPU stand for?

Unit 1　What Is a Computer?

2. Can you see the CPU from the outside of a computer?
   _____
3. When were the first CPUs used?
   _____
4. What helps to lower the prices of personal computers?
   _____
5. Why is the CPU the most expensive part of a computer?
   _____

Notes

supercomputer 超级计算机：运算速度极快，存储容量极大，处理能力极强，但其体积庞大，所占空间较大。

desktop computer 台式机：如今台式机较为普及，已被广泛使用。它的体积相对较小，可以置于桌面，其不同的部分由电缆或Wi-Fi连接。台式机的价格便宜，大部分家庭都能买得起。

notebook computer 笔记本电脑：笔记本电脑也被称为膝上电脑(laptop computer)。它体积较小，便于携带，但通常比台式机贵。

tablet personal computer 平板电脑：通常简写成 tablet，小巧、轻便、易携带，且应用简便，主要通过手指触碰、外接键盘/鼠标、触控笔控制来操作，可以浏览网页、看电子书、玩游戏、听音乐、看电影和聊天等。

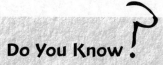

世界上第一台电子数字计算机

　　ENIAC（Electronic Numerical Integrator And Computer，电子数字积分计算机）常常被认为是世界上第一台电子数字计算机，它是全球首先采用电子技术实现的数字计算机，但因没有采用二进制数操作和存储程序控制，所以还不具备现代电子计算机的主要特征。ENIAC

是在1946年2月由美国宾夕法尼亚大学莫尔学院研制成功的。它有18 800余只电子管，运算速度达到当时继电器式计算机的1 000倍。

冯·诺依曼计算机

根据冯·诺依曼原理设计的计算机，由控制器、运算器、存储器、输入输出设备和电源五部分组成。冯·诺依曼原理的重要思想是他在1946年提出的"存储和程序"原理。世界上第一台二进制数现代电子数字计算机 EDSAC 于1949年制成，它就是按照冯·诺依曼原理研制的。我们现在使用的电脑大都是冯·诺依曼结构的计算机。冯·诺依曼 [John von Neumann（1903-1957）]，美籍匈牙利数学家、计算机科学家、物理学家，被后人称为"现代计算机之父"。

Language Practice

I. Fill in the blanks with the missing letters of the words according to the pictures.

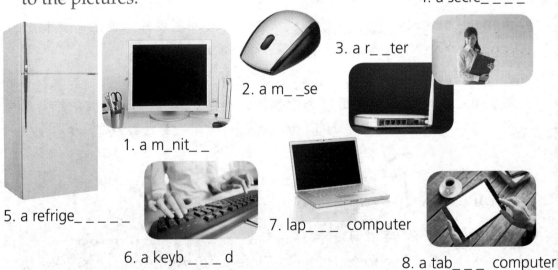

1. a m_nit_ _
2. a m_ _se
3. a r_ _ter
4. a secre_ _ _ _
5. a refrige_ _ _ _ _
6. a keyb _ _ _ d
7. lap_ _ _ computer
8. a tab_ _ _ computer

II. Complete the sentences with the words practiced above.
1. The _____ can display information on a screen.

Unit 1 What Is a Computer?

2. _____ were invented to transfer data between computer networks.
3. Mary is working as a _____ in a small computer company.
4. In summer, if you do not put food in the _____, it easily goes bad.
5. The tablet has an on-screen _____.
6. _____ computers have neither mouses nor keyboards.
7. A _____ can help us move freely on computer screens.
8. _____ computers are small enough to be put on your lap or in your briefcase.

III. Complete the passage according to the pictures.

This is my new _____ . Its colour is _____ and silver. It has a nice _____ and a new type of _____ .

The _____ also looks nice and cool. I often carry it in my bag. Sometimes I use it to do my homework, and sometimes I use it to listen to some _____ , watch movies and play games with my friends.

9

IV. Sentence completion.

1. Notebook (laptop) computers can be _____. (笔记本电脑可以四处携带。)

2. You can find _____ around you, _____ those in microwave ovens, washing machines and refrigerators. (你可以在你周围发现微型电脑，如在微波炉、洗衣机和电冰箱中。)

3. Computers are useful in _____. (电脑在很多领域都很有用。)

4. Computers can help people _____. (电脑可以帮助人们结账。)

5. People may _____ friends online. (人们可以在网上和朋友交流。)

Writing

Fill in the blanks with the information you have got from the texts.

There are _____ types of computers. They are _____, _____, _____ and _____. Supercomputers are very big and _____. Desktop computers are very popular and _____ used. _____ are also called laptop computers. Tablets have neither mouses nor _____.

Unit 1 What Is a Computer?

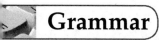 # Grammar

Tense (时态) (1)		
Simple Present	do does	Students **have** a computer class every Tuesday. The letters of CPU **stand for** Central Processing Unit. The e-classroom **makes** learning easier and more fun.
Simple Past	did	In the past, computers **were** very expensive. Peter **learned** how to use a computer ten years ago.
Simple Future	shall do will do be going to do	We **shall begin** to set up a new language lab next month. Computers **will continue** to change our lives in many more amazing ways. A newer, faster CPU **will** often **enable** you to be a better user. He **is going to study** computer science at college.

常用不规则动词的现在分词与过去式的变化形式

原形	现在分词	过去式
be	being	was / were
come	coming	came
get	getting	got
go	going	went
give	giving	gave
meet	meeting	met
rise	rising	rose
buy	buying	bought

11

I. Multiple choices.
1. I _____ to the computer lab for my research paper almost every day.
   A. went   B. shall go   C. go
2. My mom _____ me a new tablet as a birthday gift next week.
   A. gave   B. will give   C. give
3. The sun _____ in the east.
   A. rises   B. rose   C. will rise
4. Lily _____ a new laptop last month.
   A. will buy   B. buy   C. bought
5. Linda _____ Tom at the chat-room at 6:00 pm every day to practice their English.
   A. is going to meet   B. met   C. meets

II. Fill in the blanks with the correct forms of the verbs in the brackets.

Decades ago, most people _____ (have) little to do with computers. Then microcomputers _____ (come) along and _____ (change) everything. Today it is easy for nearly everyone to use a computer. Writers _____ (write), artists draw, teachers teach and students learn — all on microcomputers. In the future, computers _____ (become) more and more popular.

Word formation — Compounding (合成法)

合成法是把两个或两个以上的词按照一定的次序排列构成新词的方法。用这种方法构成的新词叫作复合词（compound）。复合词是由两个或以上词素连在一起构成的一个整体单位，是表达一个单独意思的词的结合体，如：keyboard，boyfriend等。

Unit 1 What Is a Computer?

III. Match the words in column A with those in column B to form compound words.

A	B
1. micro	a. computer
2. washing	b. wave
3. super	c. phone
4. note	d. board
5. brief	e. top
6. smart	f. case
7. key	g. book
8. lap	h. machine

Game

Across

1. Sam's dad bought a new _____ computer for him as the birthday gift.
2. You need to connect all these _____s before you turn on the computer.
3. Watch the _____, you can see what you typed in just now.
4. Linda works as a _____ to the manager in a computer company.
5. Nowadays, computers are so inexpensive that almost every family can _____ one.

Down

1. Your computer is out-of-date now. You need to _____ it to a new one.
2. It is a great fun to _____ the Internet.
3. For some people, a computer is only a game _____. But it has more functions.
4. Tablets are very _____ nowadays, especially for young people.
5. _____s were invented to transfer data between computer networks.

 Fun Time

Tech Support

A woman called the Canon help desk for a problem with her printer.

The tech asked her if she was "running it under Windows".

The woman then responded, "No, my desk is next to the door. But that is a good point. The man sitting next to me is under a window, and his is working fine."

 Project

Communicate with others!

Step 1: Within each group, talk about different types of computers.

Step 2: Share with your group members the most useful type of computers you think and explain reasons.

Step 3: Ask one classmate from each group to give an oral presentation.

Self-checklist

根据实际情况，从A、B、C、D中选择合适的答案：A代表你能很好地完成该任务；B代表你基本上可以完成该任务；C代表你完成该任务有困难；D代表你不能完成该任务。

A B C D
□ □ □ □ 1. 能掌握并能运用本单元所学重点句型、词汇和短语。
□ □ □ □ 2. 能理解并正确模仿听说部分的句子，正确掌握发音及语调。
□ □ □ □ 3. 能模仿句型进行简单的对话。
□ □ □ □ 4. 能读懂本课的短文，并正确回答相关问题。
□ □ □ □ 5. 能掌握一般现在时、一般过去时和一般将来时的用法。
□ □ □ □ 6. 能掌握基本动词过去式的特殊变化。
□ □ □ □ 7. 能掌握复合构词法。
□ □ □ □ 8. 能用课文中学习的词组造句。
□ □ □ □ 9. 能向同学介绍电脑及其种类等相关的基本知识。

Unit 2

Hardware

Unit Goals

In this unit, you will be able to
- understand listening materials about computer hardware;
- talk about hardware devices;
- understand articles about computer hardware;
- write a short paragraph about parts inside a computer box;
- demonstrate your knowledge about hardware.

Lead-in

Picture matching.

Input devices

1. mouse   2. touch screen   3. scanner   4. digital pen   5. keyboard

a

b

c

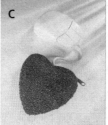

d

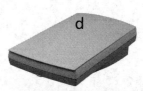

e

Unit 2 Hardware

Output devices

1. projector 2. printer 3. monitor 4. speakers

Listening and Speaking

Listen and complete.

a. Why is this in a different colour?
b. They are the parts that you can see and touch.
c. They pass information out of the computer system.
d. They are input devices, output devices and backing storage devices.
e. How many types of hardware are there?

Computer Hardware

Some students are visiting a computer company, and George, a computer programmer, is showing them around. The students are asking him some questions.

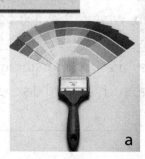

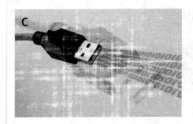

1 Student 1: Excuse me. What is hardware?
   George: Well, it's used to describe computer devices. _____ Some parts can be seen, but some parts cannot be seen from the outside. They are inside computers.

2 Student 2: _____
   George: That's a good question. There are three.
   Student 2: Then, what are they?
   George: _____

3 Student 3: That's interesting. Could you tell us more?
   George: Sure. I will draw a picture to show you. Look, here. Input devices pass information into the computer system.
   Student 3: _____
   George: Because this is not a part of hardware. All of the important things are done here.
   Student 3: And then, what is the next box?
   George: They are output devices. _____ And the last. They are backing storage devices that store programs and data, such as hard disks and USB flash disks.

Unit 2 Hardware

Role Play

A: Excuse me, Miss / Mrs / Mr ... What is / are hardware / input devices / output devices / backing storage devices?

B: Well, it is / they are ...

Words & Expressions

touch /tʌtʃ/ *n. & v.* 触摸；联系
scanner /ˈskænə(r)/ *n.* 扫描器，扫描仪
projector /prəˈdʒektə(r)/ *n.* 投影仪
input /ˈɪnpʊt/ *n. & v.* 输入
device /dɪˈvaɪs/ *n.* 装置，设备
output /ˈaʊtpʊt/ *n. & v.* 输出

storage /ˈstɔːrɪdʒ/ *n.* 存储；贮藏库
data /ˈdeɪtə/ *n.* (datum 的复数) 资料，数据

show ... around 带……参观
USB flash disk U盘

Reading

Text A

Pre-reading activities.

1. Have you heard of the following computer parts? Talk about their main functions with your partner.

| motherboard | processor | memory | bus |

2. Read the text briefly and answer the following questions.
1) What is the text mainly about?
2) What are the four hardware parts mentioned in the text?
3) What are the "buses" in a computer?

19

Reading Strategy

略读 Skimming 又称为跳读或浏览，指的是一种快速阅读文章以了解其内容大意的阅读策略。通常，阅读者有选择地进行阅读，可跳过某些细节，以求抓住文章的大意或作者的写作意图，从而加快阅读速度。

Inside a Computer

Inside a computer there is a circuit board called motherboard, and there are some hardware parts that are attached to this board.

Processor

The processor is the most important part of a computer. It is also called CPU (Central Processing Unit) or microprocessor. It is very small, but it is the "brain" of a computer system. It runs the programs, processes data, and then turns them into information that people need.

Memory

Memory stores programs and data. There are two kinds of memory — ROM and RAM. ROM stands for Read-only Memory. The processor can read from ROM but cannot write to it. And RAM stands for Random Access Memory. The processor can read from and write to RAM. There is always much more RAM than ROM.

Buses

They are not the things that can take you to and from school, but are sets of cables. They connect components on the motherboard and other hardware devices.

20

Words & Expressions

circuit /ˈsɜːkɪt/ n. 电路
motherboard /ˈmʌðəbɔːd/ n. 主板，母板
attach /əˈtætʃ/ v. 附上，贴上
processor /ˈprəʊsesə(r)/ n. 处理机，处理器
microprocessor /ˌmaɪkrəʊˈprəʊsesə(r)/ n. 微处理器
read-only /ˈriːdˈəʊnli/ n. 只读 adj. 只读的
random /ˈrændəm/ adj. 随机的，任意的
access /ˈækses/ n. 通路；入门 v. 存取；接近
bus /bʌs/ n.（计算机系统的）总线
component /kəmˈpəʊnənt/ n. 元件，部件；成分
turn ... into 把……转变成

True or false.

1. The motherboard is inside a computer.
2. CPU can process information and turn it into data.
3. The microprocessor is the "brain" of a computer system.
4. ROM always holds much more data than RAM.
5. The word "bus" has a different meaning in computer English.

Text B

Pre-reading questions.

1. Do you know anything about some "firsts" in computer history? Share your information with your classmates.
2. List some of the latest models of hardware you know or use.

History of Computer Hardware

Since the early twentieth century, the development of computers has seen numerous changes in the hardware. Here are some of the examples.

Mouse

In 1963 Douglas Engelbart invented the first mouse, which had two wheels set at a 90-degree angle to each other to keep track of the movement. The ball mouse was not invented until 1972. And the optical mouse was invented around 1980. It did not become popular until much later.

Laser Printer

Gary Starkweather invented the first laser printer in 1969. However it received wide use only after IBM introduced their branded laser printer known as IBM 3800 in 1976. It was as large as a room.

Floppy Disks

Floppy disks were invented in 1970 and used till 1990s. However, today their use has become history.

Web Server

The first web server was born in 1991. It was a NeXT workstation that Tim Berners-Lee used when he invented the World Wide Web at CERN. A note on the computer said, "This machine is a server. DO NOT POWER IT DOWN!!" It meant that if you had shut it down in the early days you would have shut down the entire WWW.

Words & Expressions

numerous /ˈnjuːmərəs/ *adj.* 许多的，无数的
angle /ˈæŋgl/ *n.* 角，角度
track /træk/ *n.* 踪迹
optical /ˈɒptɪkl/ *adj.* 光学的；眼的，视力的
laser /ˈleɪzə(r)/ *n.* 激光
introduce /ˌɪntrəˈdjuːs/ *v.* 介绍；传入，引进

branded /ˈbrændɪd/ *adj.* 属于品牌的
server /ˈsɜːvə(r)/ *n.* 服务器
workstation /ˈwɜːksteɪʃn/ *n.* 工作站
keep track of 跟上……的
known as 被称为；以……著称
as ... as 和……一样，像……一样
shut ... down （使机器等）关闭

Unit 2 Hardware

Short-answer questions.

Answer the following questions according to Text B.

1. What did Douglas Engelbart's first mouse look like?

   _____

2. When was the optical mouse invented?

   _____

3. Which company contributed (贡献) to the wide use of the laser printer?

   _____

4. Are floppy disks widely used today?

   _____

5. What special caution (警告) was given to the first web server?

   _____

Notes

hardware 硬件：是计算机系统的物理组成部分，包括任何外围设备，诸如打印机、路由器和鼠标。它指可以实际触摸到的对象，如硬盘、显示器、键盘、打印机和芯片等。

CPU Central Processing Unit 中央处理器：有时也称微处理器。它是电脑的核心部分，相当于电脑的大脑。

motherboard 主板：又称母板，是微型计算机的主电路板，通常含有CPU、存储器和总线，以及控制标准的外围设备（如显

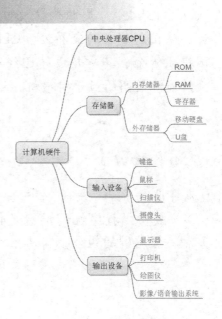

23

示器、键盘和磁盘驱动器）所需的全部控制器。

memory 存储器：计算机硬件中的存储器分为外存储器与内存储器。外存储器包括移动硬盘、U盘等；内存储器是计算机中的内部存储区域，是指以芯片形式存在的数据存储器。存储器有若干不同的类型：随机存储器(RAM)、只读存储器(ROM)、可编程只读存储器(PROM)、可擦可编程只读存储器(EPROM)、电可擦可编程只读存储器(EEPROM)等。

RAM 随机存储器：使用者可以向RAM中写数据，也可以从RAM中读数据。大多数RAM都有易失性，一旦电源关闭，它里面的任何数据都会丢失。

ROM 只读存储器：它与RAM不同，只允许读出数据，不能写入数据。

bus 总线：是将数据从计算机的一部分传到另一部分的一组连线，是计算机内部传输信息的公路。每种微处理机都具有三组总线，分别供数据、地址和控制信号使用。信息可从多个源部件中的任何一个经总线传送到多个目标部件中的任意一个。

IBM IBM 公司：即国际商业机器公司 (International Business Machines Corporation)，1911年创立于美国。在过去的一百多年里，IBM始终以超前的技术、出色的管理和独树一帜的产品引领全球信息产业的发展。

CERN European Organisation for Nuclear Research 欧洲粒子物理研究所：成立于1954年，是世界最大的粒子物理研究中心。CERN位于日内瓦的西北部，法国和瑞士的交界处。CERN是欧洲第一个联合研究机构，由20个成员国提供资金，其卓越成绩已经成为国际合作的典范。CERN目前拥有世界上设计能量最高的粒子对撞机，还是互联网的诞生地。

Do You Know？

计算机键盘

最初，打字机的键盘是按照字母顺序排列的，但如果打字速度过快，某些键的组合很容易出现卡键问题，于是克里斯托夫·拉森·施奥莱(Christopher Latham Sholes)发明了QWERTY键盘布局，他将最常用的几个字母安置在相反方向，最大限度放慢敲键速度以避免卡键。施奥莱在1868年申请专利，1873年使用此布局的第一台商用打字机成功投放市场。今天的计算机键盘就是在此基础上演变而来的。

计算机硬件使用禁忌

在使用各种电脑硬件的过程中应遵循一定的注意事项，以维持其正常的工作状态，延长其使用寿命。但每种硬件的使用禁忌不尽相同。硬盘最忌震动，主板最忌静电和形变，CPU最忌高温和高电压，内存最忌超频，U盘忌讳读写数据时拔出，鼠标最忌灰尘、强光和拉拽，键盘最忌潮气、灰尘以及拉拽，电源最忌反复开机关机，而显示器最忌冲击、高温、高压、灰尘、高亮度、高对比度、电子灼伤等。

Language Practice

I. Fill in the blanks with the missing letters of the words according to the pictures.

1. k_ _ b_ _rd

2. di_ _tal pen

3. pr_ nt_ _

6. micropr_ce_ _ or

5. t _ _ ch scr _ _ n

4. sca _ _ er

7. sp _ _ ker

8. l_ s _ _

II. Complete the sentences with the words practiced above.

1. A _____ allows the tablet user to input easily.
2. Mary has decided to buy a Bluetooth _____.
3. In many public places there are some _____ instead of keyboards.
4. The computer _____ was developed from the typewriter.
5. _____ can be used to input pictures or texts into a computer.
6. Carol is trying to find a _____ to get the pictures printed out.
7. A _____ is very small but very powerful.
8. Scientists use _____ beams (光束) to measure the distance between the Earth and the Moon.

Unit 2 Hardware

III. Complete the passage with the words in the box.

| data | output | hardware | back-up | input |

Computer systems are made up of _____ and software. They process _____, which are given to them as _____. They give the results of the processing back to users as _____. They can store data for people to use later on _____ storage.

IV. Sentence completion.

1. Many computer hardware devices _____ the motherboard. (电脑的很多硬件设备都是附着在主板上面的。)
2. This software can _____ your mobile phone _____ a scanner. (这个软件可以将你的手机变为扫描仪。)
3. The laser printer was not _____ 1976. (直到1976年激光打印机才被广泛使用。)
4. The first computer was _____ a room, but today's computers are much smaller. (最早的电脑像一间屋子那么大，而今天的电脑就小得多了。)
5. Charles Babbage was _____ the Father of Computers. (查尔斯·巴比奇被称为"计算机之父"。)

Writing

Fill in the blanks with the information you have got from the texts.

There are many hardware parts inside a computer box, such as _____, _____ and _____. They are mostly attached to the _____. The most important part is the processor, or _____, serving as the "brain" of a computer system. RAM and _____ are two kinds of memory, which

_____ programs and data. And buses _____ components on the motherboard with other hardware devices.

Grammar

Tense (时态) (2)		
Present Progressive	am doing are doing is doing	I **am working** on my computer. Some students **are visiting** a computer company, and George, a computer programmer, **is showing** them around.
Past Progressive	was doing were doing	I **was composing** some music with the software when you called. They **were having** lessons about computer software this time yesterday.
Present Perfect	have done has done	We **have not solved** the software problem yet. Julia **has** just **connected** some hardware to her computer.
Past Perfect	had done	I **had finished** the book on iOS before your last visit.

I. Multiple choices.

1. When I saw her, she _____ at the computer.
   A. worked   B. has worked   C. was working
2. Nowadays touch screens _____ increasingly common in our lives.
   A. are becoming   B. were becoming   C. became
3. Great changes _____ place in computer industry in the past ten years.
   A. took   B. have taken   C. are taken
4. Don't shut down the power while the system _____.
   A. has upgraded   B. upgrades   C. is upgrading

5. He _____ some basic training before he worked as a secretary.
   A. was received   B. had received   C. has received

II. Fill in the blanks with the correct forms of the verbs in the brackets.

Toshiba _____ (introduce) flash memory in the early 1980s to replace existing data-storage media such as magnetic tapes and floppy disks. Since then, its use _____ (expand) to digital cameras, smartphones, and MP3 players. As its cost _____ (become) cheaper in the early 21st century, flash memory also _____ (begin) appearing as the hard disk in laptop computers.

Word formation — Conversion (转化法)

转化法是指在英语构词时把一种词性用作另一种词性而词形不变的方法。其中，某些词在具体使用时可以体现动词与名词之间的转化，其前后词义通常有着密切的联系，但有时差异也很大。例如：

I *try* to follow his steps. (v. 试图；努力)

Let us have a *try*. (n. 尝试)

III. Read the following sentences and tell the Chinese meanings of the underlined words.

1. 1) The two cables are <u>touching</u> each other.
   2) I gave a <u>touch</u> on the screen.
2. 1) A mouse is a popular <u>input</u> device.
   2) Now please <u>input</u> your name and password.
3. 1) The only <u>access</u> to a skilful programmer is by working hard.
   2) She <u>accessed</u> three different files to find the right information.
4. 1) How many movies can you <u>store</u> in this hard disk?
   2) The <u>store</u> sells many computer components.

Game

Across

1. CPU is the most important _____ of a computer system.
2. My computer suddenly made noises. Maybe something went wrong with the _____.
3. Please double check the computer and make sure it is _____ed off.
4. A scanner is an optic recognition _____ used for direct computer data input.
5. The Internet has _____ed the whole world into a global village.

Down

1. Input and _____ devices enable people and computers to communicate.
2. The computer technician had a good knowledge of _____.
3. Last week HUAWEI _____d its new laptop to the market.
4. The WWW system is based on a standard client-_____ model.
5. In a computer, the _____ is the part where information is stored.

 Fun Time

Mouses (Mice)

The computer company I worked for placed an order for computer mouses (mice) from Japan. After the delivery (交付) period had passed, we called the airport to ask about what had happened to the goods (货物). The official said that it was nowhere to be found and should be

Unit 2 Hardware

reported as missing. Some time later he called us back to say that the package had been found. When we asked where it had been, he replied shyly, "In quarantine (检疫处)."

Project

Communicate with others!
Step 1: Divide the class into 6 groups. Ask 3 groups to list common input devices and discuss when they're usually used.
Step 2: Ask another 3 groups to list common output devices and discuss when they're usually used.
Step 3: Invite some students to report the results.

Self-checklist

根据实际情况,从A、B、C、D中选择合适的答案:A代表你能很好地完成该任务;B代表你基本上可以完成该任务;C代表你完成该任务有困难;D代表你不能完成该任务。

A B C D

☐ ☐ ☐ ☐ 1. 能掌握并能运用本单元所学重点句型、词汇和短语。

☐ ☐ ☐ ☐ 2. 能理解并正确模仿听说部分的句子,正确掌握发音及语调。

☐ ☐ ☐ ☐ 3. 能模仿句型进行简单的对话。

☐ ☐ ☐ ☐ 4. 能读懂本课的短文,并正确回答相关问题。

☐ ☐ ☐ ☐ 5. 能掌握现在进行时、过去进行时的用法。

☐ ☐ ☐ ☐ 6. 能了解转化构词法中动词与名词之间的相互转化。

☐ ☐ ☐ ☐ 7. 能用课文中学习的词组造句。

☐ ☐ ☐ ☐ 8. 能向同学介绍电脑硬件及其功能等相关基本知识。

Unit 3 Software

Unit Goals

In this unit, you will be able to
- understand listening materials about different operating systems;
- talk about operating system preferences;
- understand articles about software and freeware;
- write a short paragraph about computer software;
- demonstrate your knowledge about software.

Lead-in

Fill in the blanks with the names of different software.

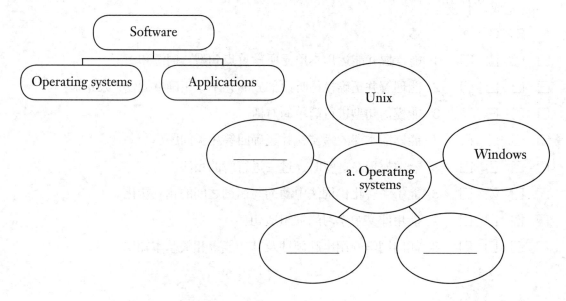

Unit 3 Software

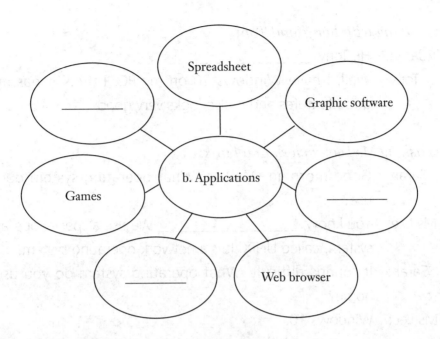

Listening and Speaking

Listen and complete.

a. I use iOS (operating system) on my iPad.
b. What operating system do you use?
c. I work for an international bank.

What Operating System Do You Use?

All computers must use operating systems. Different people have different likes and dislikes about the operating systems they use. Sarah, a computer school student, is asking people the question.

33

She is talking with her friend Tony.

1 Sarah: Hi, Tony. _____

Tony: Well, I have Windows 10 on my PC. I think it has many good features and it also looks very good.

She is asking Ms Lee, a clerk in a bank.

2 Sarah: Good morning, Ms Lee. What operating system do you use?

Ms Lee: You know, _____ We use a special operating system, called UNIX. It is a network operating system.

Sarah: It sounds difficult. What operating system do you use at home?

Ms Lee: Windows 10.

She is talking with Mr Wang, a sales manager of IBM, who is using his iPad.

3 Sarah: Excuse me, sir. What operating system do you use?

Mr Wang: _____ It was already installed when I bought it. It serves as a good assistant for me.

Role Play

A: What operating system do you use at home / at school / at work?

B: Well, I use Windows 10 / Mac OS / iOS / Android / Linux ...

Unit 3 Software

Words & Expressions

operate /'ɒpəreɪt/ v. 操作，开动；动手术
spreadsheet /'spredʃiːt/ n. 电子制表软件；电子数据表
graphic /'ɡræfɪk/ adj. 绘画的，图解的
presentation /ˌprezn'teɪʃn/ n. 介绍；陈述
browser /'braʊzə(r)/ n. 浏览器
application /ˌæplɪ'keɪʃn/ n. 应用程序；应用；申请
assistant /ə'sɪstənt/ n. 助手；助教
install /ɪn'stɔːl/ v. 安装

Text A

Pre-reading activities.

1. Name two or three kinds of system software you know.

2. Name two or three kinds of application software you use. Compare your answers with those of your peers.

3. Read the text silently and answer the following questions. Note down the time you have used.
   1) What is software?
   2) What is system software in charge of?
   3) What is the most common application software?

Reading Strategy

默读 Silent Reading 默读也是一种阅读技巧，它要求阅读者尽量扩大视幅，对篇章的文字符号整句整行进行识别，从而避免出声或在心里逐字逐句阅读。与传统阅读比较，默读的速度更快、理解更深入，在查阅文件资料、阅读报纸或杂志等情况时有很大帮助。

35

Software

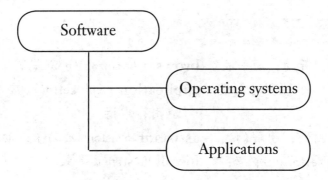

Software refers to the programs that our computer systems run on. There are two types of software — operating systems and applications.

Operating Systems

All computers use operating systems. System software controls computer hardware, and loads software into RAM. It is also called background software. It is in charge of running programs, storing data and programs, and processing data. Windows 10 is one of the best known examples.

Applications

Application software is sometimes called applications or end-user software. It performs real-world jobs that people do in their daily life, work and study. They use software to create files, manage personal finance, make movies, surf the Internet, download music from the Internet, develop a website and play computer games. Some of the most commonly used application programs are word processors, database systems, presentation systems, spreadsheet programs, and desktop publishing programs.

Unit 3 Software

Words & Expressions

refer /rɪˈfɜː(r)/ v. 提到，涉及
background /ˈbækɡraʊnd/ n. 背景，后台
end-user /ˌendˈjuːzə(r)/ n. 终端用户
perform /pəˈfɔːm/ v. 执行；表演
finance /faɪˈnæns/ n. 财政，金融
download /ˌdaʊnˈləʊd/ v. 下载

website /ˈwebsaɪt/ n. 网站
database /ˈdeɪtəbeɪs/ n. 数据库，资料库
publish /ˈpʌblɪʃ/ v. 出版，刊印

refer to 指的是，说的是
in charge of 负责，主管

True or false.
1. Software is not devices but programs.
2. The two basic types of computer software are programs and applications.
3. All programs are designed for end-users.
4. Word processor is an example of system software.
5. Operating systems control computer hardware.

Text B

Pre-reading questions.
1. Have you tried some free software? If yes, name one or two of them.
2. Discuss with your deskmate about the differences between freeware and pirated software.

Freeware

Freeware refers to the software that we can use and share with others free of charge. The opposite of freeware is, of course, payware.

Many designers would like to make their software free and open to the public. Why? Usually they have their own reasons. Some college professors and students enjoy making academic advances in computer science. Some people develop freeware just for fun or for proving their programming

37

skills. Also some people want to test the marketability of their software. If the freeware turns out to be popular, they will sell the program for profit after the trial period (often one to three months) expires.

One thing to be noted is that pirated software is not freeware, as it is made to be free and costless against the will of the license holder.

Words & Expressions

freeware /ˈfriːweə(r)/ *n.* 免费软件
opposite /ˈɒpəzɪt/ *n.* 对立物，相反的事物 *adj.* 对面的；对立的，相反的
designer /dɪˈzaɪnə(r)/ *n.* 设计师
academic /ˌækəˈdemɪk/ *adj.* 学术上的
advance /ədˈvɑːns/ *n. & v.* 前进；进步，发展
marketability /ˌmɑːkɪtəˈbɪləti/ *n.* 可销售性，市场性

trial /ˈtraɪəl/ *n.* 试用
expire /ɪkˈspaɪə(r)/ *v.* 期满，终止
pirated /ˈpaɪrɪtɪd/ *adj.* 盗版的
license /ˈlaɪsns/ *n.* 许可证，执照
share with 与……分享
free of charge 免费
turn out 结果是，被证明是
trial period 试用期
license holder 许可证持有者

Short-answer questions.

Answer the following questions according to Text B.

1. What is freeware?

___

2. What is the opposite of freeware?

___

3. Do all the designers develop freeware for free?
   _____

4. How long does the trial period of a freeware usually last?
   _____

5. Why is pirated software not freeware?
   _____

Notes

UNIX Operating System UNIX操作系统：由美国电报电话公司的Bell实验室于1969年开发，最初配置在DEC公司的PDP小型机上。UNIX操作系统是唯一一个可以在微机工作站、小型机、大型机上都能运行的操作系统，也是当今世界流行的多用户、多任务操作系统。

LINUX Operating System LINUX操作系统：LINUX是目前全球最大的自由软件。简单地说，它是一套类似UNIX、可以免费使用和自由传播的操作系统。LINUX继承了UNIX以网络为核心的设计思想，具有良好的开放性，是一个性能稳定的多用户网络操作系统，主要用于基于Intel x86系列CPU的计算机上。这个系统是由世界各地成千上万的程序员设计和实现的，其目的是建立不受任何商品化软件的版权制约的、全世界都能自由使用的UNIX兼容产品。

Novell Operating System Novell操作系统：Novell被认为是网络服务器操作系统的鼻祖，其产品早在20世纪80年代就进入中国。Novell公司是全球著名的软件公司，曾经在全世界软件行业中排名第五。

Android Operating System Android 操作系统：是一种基于Linux的自由及开放源代码的操作系统，主要使用于移动设备，如智能手机和平板电脑，由Google公司和开放手机联盟领导及开发。Android的开放性优势使其拥有最多的app开发者及数以千万计的app。

iOS Operating System iOS操作系统：是由苹果公司开发的移动操作系统。苹果公司最早于2007年1月9日的Macworld大会上公布这个系统，最初是设计供iPhone使用的，后来陆续套用到iPod touch、iPad以及Apple TV等产品上。iOS与苹果的macOS操作系统一样，属于类Unix的商业操作系统。iOS平台拥有数量庞大的移动app，几乎每类app都有数千款。

MacOS Catalina Operating System Mac OS X操作系统：是苹果电脑操作系统软件。它是全球领先的操作系统，基于UNIX基础，设计简单直观，且功能强大、高度兼容，还提供超强性能、超炫图形，并支持互联网标准。

Windows 10 Operating System Windows 10 操作系统：是由微软公司开发的应用于计算机和平板电脑的操作系统，于2015年7月29日发布正式版。Windows 10 操作系统在易用性和安全性方面有了极大的提升，除了针对云服务、智能移动设备、自然人机交互、虚拟现实等新技术进行融合外，还对固态硬盘、生物识别、高分辨率屏幕等硬件进行了优化完善与支持。

Do You Know?

操作系统

操作系统是计算机运行最重要的程序，是用户和系统的界面，每台通用计算机都必须有一个操作系统来运行其他的程序，它提供一个软件平台。操作系统的卓越工作，既能保证系统资源的充分利用，又能使用户方便地使用计算机。操作系统为使用者提供一个使用计算机和应用软件的界面，这就是人们常说的用户界面。用户界面可分为三种类型：命令式用户界面、菜单式用户界面和最简单常用的图形式用户界面（GUI）。使用图形式用户界面，只需点击桌面上的图形即可。

Unit 3 Software

程序员
程序员是设计和编写软件的专业人员。他们在编写软件时使用特殊的语言，常用的编程语言有Visual Basic、C、C++、Java和Python。

Python语言
Python是一种跨平台的计算机程序设计语言。它是一种面向对象的动态类型语言，最初被设计用于编写自动化脚本（shell），随着版本的不断更新和语言新功能的添加，已经被用于独立的、大型项目的开发上，越来越多的学校也采用Python来教授程序设计课程。Python极其容易上手，而且免费、开源，有着越来越丰富的功能库，成为近些年来最热门的编程语言之一。

Language Practice

I. Fill in the blanks with the missing letters of the words according to the pictures.

1. op_r_ting s_st_m

2. _ppl_cat_ _ns

3. downl_ _d songs

4. c_ _p_se music

5. pr_s_ _tat_ _ _

6. d_t_b_se

7. w_bs_te

8. f_rew_ll

II. Complete the sentences with the words practiced above.
1. The _____ of our company was attacked yesterday by a hacker.
2. _____ is an effective measure of network security.
3. It is very convenient to manage the salary of the employees with a _____ system.
4. Sam reported his business plan with a PPT _____.
5. Is it illegal to _____ for free from the Internet?
6. The _____ I use in my computer is Windows 10.
7. Those programmers are making great efforts to design better _____.
8. Is there software to help _____?

III. Complete the passage with the words in the box.

speaker   software   hardware   application   memory

Denver has just bought a new computer. It has all the best _____. The CPU is the latest Intel one with a big capacity of _____, a huge monitor and nice _____. But it only comes with an operating system

— Windows 10 and has no _____ yet. He decides to buy some this weekend. He is going to buy a _____ suite (套件) like *Microsoft Office* and some special software for computer games.

Ⅳ. Sentence completion.

1. You can _____ your stories _____ friends on either Weibo or WeChat. (你可以在微博或微信上和朋友们分享你的故事。)
2. Who is _____ software development in your company? (你们公司谁负责软件开发？)
3. The new operating system _____ to be very successful. (新的操作系统被证明非常成功。)
4. Application software can help the user to better _____. (应用软件可以帮助使用者更好地履行他/她的任务。)
5. The firewall software will no longer be free when _____ expires. (试用期一过，这个防火墙软件就不再免费了。)

Writing

Fill in the blanks with the information you have got from the texts.

There are _____ types of software. One is _____, mainly in charge of _____ programs, storing data and programs and _____ data. The other is called _____ or end-user software, enabling people to _____ real-world tasks in daily work and study. For _____, word processors and _____ programs are the most commonly used application programs.

Grammar

Gerund (动名词)		
subject		**Programming** is the act of setting out a sequence of instructions.
object	of verb	Stop **tapping** the keys, please. Some college professors and students enjoy **making** academic advances in computer science.
	of a phrasal verb	Simon has given up **playing** computer games.
	of adjective	The engineers are busy **testing** the speakers.
	of preposition	System software is in charge of **running** programs, **storing** data and programs, and **processing** data.
predicative		Belinda's job is **checking** the files.
attribute		There is a big **swimming** pool in the yard of the computer center. We use a special **operating** system called UNIX.
complement		Mr Smith insists on us **downloading** the software.

I. Multiple choices.

1. Greg enjoys _____ friends online and _____ with them.
   A. making, chatting   B. to make, to chat   C. making, to chat

2. How can you keep the machine _____ when you are away?
   A. run   B. being run   C. running

3. _____ a software can be very demanding.
   A. Develop   B. Developing   C. Having developed

4. Today people can do their banking online or on their smartphones without _____ the houses.
   A. left   B. to leave   C. leaving

5. Python is one of the most useful _____ languages.
   A. programmed   B. programming   C. program

II. Fill in the blanks with the correct forms of the verbs in the brackets.

_____ (write) letters, memos (备忘录) or reports are ways most people use computers. With word _____ (process) software or a word processor, users can easily create and edit their words and text on a computer screen. They have no trouble in _____ (make) changes, or worrying about _____ (retype) the entire document if they make any mistakes. Even their spelling or grammatical mistakes can be _____ (check) out and corrected.

Word formation — Conversion (转化法)

在转化构词法中，某些词语也可以体现形容词与名词之间的互换。例如，He is very *rich* (adj. 富有的). He stole from the *rich* (n. 有钱人，富人) to give to the poor.

III. Read the following sentences and tell the Chinese meanings of the underlined words.
1. 1) The speakers are put at the opposite ends of the desk.
   2) ENIAC was rather heavy, but a laptop is just the opposite.
2. 1) Smoking is not allowed in public places like schools or hospitals.
   2) The freeware does good to the public.
3. 1) He does not know the difference between right and wrong.
   2) Something goes wrong with my new printer.
4. 1) Most people love to live in peace and quiet.
   2) Can you keep the children quiet? I am trying to write a report.

Game

Across

1. Users can apply spreadsheets to create and manage _____s.
2. You had better pay close attention to viruses when _____ing new software.
3. The system can only _____ on a local basis.
4. Software _____ requires much patience and hard work.
5. Students _____ a lot from open educational resources online.

Down

1. Everybody can _____ music from iTunes Store with an Apple ID.
2. _____ing the Internet has become an important part of our daily life.
3. How much is the _____ for reading this file?
4. Tablets serve as good _____s for many business people.
5. Mr Wang is a software expert in _____. He can help better handle your money.

 Fun Time

Programmers' Wisdom

Every program has at least one bug and can be shortened by at least one instruction — from which, by induction (归纳), one can deduce (推论) that every program can be reduced to one

46

instruction which does not work.

Project

Communicate with others!

Step 1: Within each group, talk about the application software you have in your computers.

Step 2: Share with your group members your favourite applications and explain why you like them.

Step 3: Ask one classmate from each group to give an oral presentation to introduce their favourite applications in class.

Self-checklist

根据实际情况，从A、B、C、D中选择合适的答案：A代表你能很好地完成该任务；B代表你基本上可以完成该任务；C代表你完成该任务有困难；D代表你不能完成该任务。

A B C D
☐ ☐ ☐ ☐ 1. 能掌握并能运用本单元所学重点句型、词汇和短语。
☐ ☐ ☐ ☐ 2. 能理解并正确模仿听说部分的句子，正确掌握发音及语调。
☐ ☐ ☐ ☐ 3. 能模仿句型进行简单的对话。
☐ ☐ ☐ ☐ 4. 能读懂本课的短文，并正确回答相关问题。
☐ ☐ ☐ ☐ 5. 能了解动名词的构成和基本用法。
☐ ☐ ☐ ☐ 6. 能了解转化构词法中名词与形容词之间的相互转化。
☐ ☐ ☐ ☐ 7. 能用课文中学习的词组造句。
☐ ☐ ☐ ☐ 8. 能向同学介绍电脑软件及其种类等相关基本知识。

Unit 4

Basic Computer Skills

Unit Goals

In this unit, you will be able to

- understand basic computer operating skills;
- talk about basic computer skills;
- understand articles about basic computer skills;
- write a short paragraph about how to acquire basic computer skills;
- demonstrate your knowledge about basic computer skills.

Lead-in

Picture matching.

An Introduction to GUI

1. scroll bar
2. window
3. pointer
4. menu
5. icon
6. GUI
7. function keys
8. dialogue box

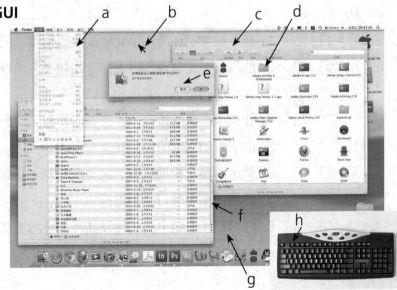

48

Unit 4 Basic Computer Skills

Listening and Speaking

Listen and complete.

a. Push the power button.
b. Click on the power button, and click on "Shut Down".
c. How can I start a GUI?
d. To shut down a GUI is simple, too.

How to Start and Shut Down the GUI?

Peter has bought a new computer. The computer has a Windows 10. But he does not know how to start and shut down the Windows 10 GUI. He is asking his computer teacher, Miss White.

❶ Peter: Excuse me, Miss White. _____
Miss White: Well, to start a GUI is as simple as turning on your computer.
Peter: You mean, just push the power button?
Miss White: Yeah, you are right.

❷ Peter: Let me try. _____
Miss White: First, the computer tests itself. Then it loads the system software.
Peter: Now I'm looking at the desktop.
Miss White: Great. The GUI has been started.

❸ Peter: Good afternoon, Miss White. How can I shut down a GUI?
Miss White: _____
Peter: Should I push the power button to shut it down?
Miss White: No, you shouldn't.

4
Peter: What should I do, Miss White?
Miss White: Open the "Start" menu.
Peter: What should I do then?
Miss White: _____
Peter: Now both the GUI and the machine are shut down. Thank you, Miss White.

Role Play

A: Excuse me, Miss White. How can I start a GUI / shut down a GUI / ...?
B: Well, it's simple / hard / easy / ...

Words & Expressions

introduction /ˌɪntrəˈdʌkʃn/ n. 介绍
scroll bar n. 滚动条
pointer /ˈpɔɪntə(r)/ n. 光标；指针
icon /ˈaɪkɒn/ n. 图标；肖像
button /ˈbʌtn/ n. 按钮；钮扣

click /klɪk/ v. 点击

GUI 图形用户界面 (Graphical User Interface，又称图形用户接口)
turn on 打开 (电视、计算机等)

Reading

Text A

Pre-reading activities.

1. Can you use a computer? If yes, describe the process of formatting a disk.

50

2. Show your classmates how to set up a new folder.

3. What do you use the following icons or function key for?

Reading Strategy

推理阅读 **Inferring** 推理阅读就是根据文章结构、逻辑关系和语篇特点等理解文章隐含之意；利用注释和逻辑手段（如归纳），借助图表等对文章各种语言信息和非语言信息作出适当的推理和判断。

Basic Skills to Use a Computer

Opening a Window

Double-click an icon on the desktop. You may open several windows one after another.

Closing a Window

On the "File" menu, single-click "Exit".

Or single-click the button ✖ in the upper-right corner.

Starting a Program

On the desktop, double-click the icon of the program you want to start.

Or (1) Single-click "Start" button.

    (2) Point to "All Programs", then single-click the program you prefer.

Formatting a USB Drive

Plug in your USB device to a free USB port.

On the desktop, double-click the icon to open the window.

Single-click the icon to select it.

Move the mouse pointer over the selected icon and right-click to display the shortcut menu.

Select "Format" to display the Format dialogue. Under "File Type", select "Full". Single-click "Start" to begin the formatting process.

Creating a Folder

On the desktop, double-click the icon .

Double-click the icon , either Drive C or Drive D.

Put the pointer in any blank space, right-click your mouse, point to "New" and select "Folder".

Type a name for the new folder.

Press "Enter".

Folders are a great help to organise our files.

Getting and Using Help

On the menu bar, click "Help".

Or press the function key .

Have you got the skills above? If yes, you are on the way to be a computer expert already.

Words & Expressions

format /ˈfɔːmæt/ v. 格式化（磁盘）
folder /ˈfəʊldə(r)/ n. 文件夹
double-click n. & v. 双击
single-click n. & v. 单击
exit /ˈeksɪt/ n. 退出；出口
　　　v. 离开当前命令；退出
upper-right adj. 右上的
port /pɔːt/ n. 端口
right-click n. & v. 右击；击右键
display /dɪˈspleɪ/ v. & n. 陈列；展览；显示
select /sɪˈlekt/ v. 选择，挑选

shortcut /ˈʃɔːtkʌt/ n. 捷径
organise /ˈɔːgənaɪz/ v. 组织
expert /ˈekspɜːt/ n. 专家
USB (Universal Serial Bus) 通用串行总线（一种数据通讯方式）
one after another 一个接一个，依次的
plug in 插入
point to 指向
on the way 在途中

True or false.
1. We can only open one window every time.
2. A small picture that represents a file or program is called an icon.
3. We can single-click an icon to open a window.
4. We can use folders to organise our files.
5. Push the power button to close windows.

Text B

Pre-reading questions.
1. What other basic computer skills do you think you need to know in addition

to those mentioned in Text A?

2. Which one are you good at? Show your classmates how to use it.

Some Other Basic Computer Skills

Nowadays, almost everyone knows how to use a computer. Besides some skills mentioned in Text A, you also need to know the following.

Using a word processor

It involves typing basically, but when using computers it is a little different from the common typewriter. The best of word processing is that you can modify with ease any part of the text you entered.

Using a spreadsheet

Spreadsheets are great for budgets, financial statements and other tasks that need calculations. Knowing the basics on how to use and read a spreadsheet will be very helpful.

Creating and using a database

One also needs to know how to create and use a simple database. There are millions of information points that one gathers. A database is an effective tool to organise and present the data received.

Creating / Using a PowerPoint

Every job involves some type of reporting. The best way to make a presentation is by using PowerPoint.

Using the Internet

Computers and the Internet today are almost the same thing. The Internet opens a window for you to the world. To use the Internet, you will need to know what a search engine is and how it works; what a keyword is and how it works; how to open a browser and how to use a link, etc.

Unit 4 Basic Computer Skills

Words & Expressions

mention /'menʃn/ v. 提及
involve /ɪn'vɒlv/ v. 包含
modify /'mɒdɪfaɪ/ v. 修改
budget /'bʌdʒɪt/ n. 预算
financial /faɪ'nænʃl/ a. 财政的，金融的
statement /'steɪtmənt/ n. 结算单，报表
calculation /ˌkælkjʊ'leɪʃn/ n. 计算

basic /'beɪsɪk/ n. (通常用复数) 基本因素，基本原理
present /prɪ'zent/ v. 呈现
with ease 轻而易举地
search engine 搜索引擎

Fill in the chart with the information in Text B.

Basic Computer Skills	Basic Functions of Each Skill
Word-processing	
Using a spreadsheet	
Creating and using a database	
Creating / Using a PowerPoint	
Using the Internet	

Notes

pointer 光标：它是在Windows下操作各种软件最方便的工具。我们常常通过在桌面上移动鼠标来移动光标进行操作，因而它又被称作鼠标指针。

⌛ 沙漏光标，表示正在运行程序，需要等待。

↖ 指示光标，用来选择窗口元素。

↔ 双向箭头光标，表示此处为可移动的边界，此时可以移动边界。

| 竖线光标，表示输入位置。

Do You Know?

Windows 10 的主要功能

生物识别：让你通过指纹识别、面部或虹膜扫描进行登入。

Cortana搜索功能：可以用它来搜索硬盘内的文件、系统设置、安装的应用、甚至互联网中的其他信息。作为一款私人助手服务，Cortana还能像在移动平台那样帮你设置基于时间和地点的备忘录。

多桌面：如果用户没有多显示器配置，但依然需要对大量的窗口进行重新排列，那么Windows 10的虚拟桌面应该可以帮到用户。

开始菜单进化：微软在Windows 10中带回了用户期盼已久的开始菜单功能，你不仅会在左侧看到系统关键设置和应用列表，标志性的动态磁贴也会出现在右侧。

任务切换器和任务栏：Windows 10的任务切换器通过大尺寸缩略图的方式对内容进行预览。在Windows 10的任务栏中，新增了Cortana和任务视图按钮，可以查看可用的Wi-Fi网络，或是对系统音量和显示器亮度进行调节。

通知中心：让用户可以方便地查看来自不同应用的通知，此外，通知中心底部还提供了一些系统功能的快捷开关，比如平板模式、便签和定位等。

文件资源管理器升级：Windows 10的文件资源管理器会在主页面上显示用户常用的文件和文件夹，让用户可以快速获取自己需要的内容。

新技术融合：在易用性、安全性等方面进行了深入的改进与优化，针对云服务、智能移动设备、自然人机交互等新技术进行融合。

Unit 4 Basic Computer Skills

Language Practice

I. Fill in the blanks with the missing letters of the words according to the pictures.

1. a d_ _ logue box

2. _c_ns

3. some f_ld_ _s

4. b_tt_ _s

5. scr_ll b_ _

6. a p_ _nter

7. win_ _ _

8. br_ _ser

II. Complete the sentences with the words practiced above.
1. Move your mouse, and the _____ on the screen moves, too.
2. Don't open too many _____ at a time, or your computer will crash.
3. You can find what is wrong with these _____.
4. Push the power _____, and the computer will be turned on.
5. A _____ in a computer can hold much more files than the one on

57

your bookshelf.

6. Double-click the _____ on the desktop, and we can start a program.

7. Dragging the _____, you may come to the end of the file easily.

8. If you have a faster _____, you will feel much more excited when you surf the Internet.

III. Rearrange the steps to create a folder.

(1) Type a name for the new folder.

(2) Right-click your mouse, point to "New" and select "Folder".

(3) Double-click the icon, either Drive C or Drive D.

(4) Press "Enter".

(5) On the desktop, double-click the icon.

IV. Discuss with your partner whether you should single-click or double-click to carry out the following steps and see who is quicker to answer.

1. To select, _____ the icon.

2. To close a window, _____ "Exit" on the "File" menu or _____ the button ✗ in the upper-right corner.

Unit 4 Basic Computer Skills

3. To start a program, _____ the icon of the program you want to start on the desktop, or _____ "Start" button, point to "All Programs", and then single-click the program you need.

4. To find a folder, _____ the icon 🪟 , either Drive C or Drive D.

5. To open a window, _____ an icon on the desktop.

6. To get help, _____ "Help" on the File menu.

7. To begin the formatting process, _____ "Start".

V. Sentence completion.

1. Press the power button, and you may _____. (按下电源键，你可以打开电脑。)

2. The monitor is not on, _____. (显示屏没亮，因为你没插电源。)

3. The students entered the computer lab _____. (学生们依次进入机房。)

4. People can get different kinds of information _____ online. (人们能从网上很容易地获取各种信息。)

5. _____, and you are on the way to be an expert. (按照这本手册去做，你离成为专家就不远了。)

Writing

Fill in the blanks with the information you have got from Text A.

To open a window: _____
To close a window: _____
To start a program: _____
To get help: _____

Grammar

\multicolumn{2}{c}{}	Infinitive (动词不定式)
subject	**To start** a GUI is as simple as turning on your computer.
object	Alice promised not **to spend** the money on software.
predicative	My son's wish is **to have** a laptop computer.
complement of the object	Tom asked me **to clean** the keyboard.
attribute	A database is an effective tool **to organise and present** the data received.
adverbial — of purpose	Select "Format" **to display** the Format dialogue.
adverbial — of result	I was too nervous **to remember** the steps to create a folder.
adverbial — of cause	Simon was happy **to see** the computer work again.

I. Multiple choices.

1. Move the mouse pointer over the selected icon and right-click _____ the shortcut menu.

    A. displaying   B. displayed   C. to display

2. Ruby's laptop crashed down and she asked me _____ her reboot it.

    A. to help   B. helping   C. help

3. Bob went to a computer shop _____ a memory stick.

    A. buying   B. to buy   C. bought

4. _____ the basics on how to use and read a spreadsheet will be very helpful.

    A. To knowing   B. To know   C. know

5. The best way _____ a presentation is by _____ PowerPoint.
   A. making ... to use   B. to make ... using   C. making ... using

II. Fill in the blanks with the correct forms of the verbs in the brackets.

_____ (manage) files on your computer is like managing books on your bookshelf. You can _____ (move) a file or copy a file from one folder to another or from one disk to another. It is like _____ (move) a book on your bookshelf to a new location. You can also delete a file from a folder or a disk. It is like moving away a book from your bookshelf. The easiest way to move, copy or delete a file is _____ (click) the right key of the mouse and work with the shortcut menu after _____ (select) the file.

Word formation — Conversion (转化法)

在转化构词法中,形容词转化为动词的现象虽然不像名词转化为动词那样常见,但形容词转化为动词时,语义比较简单,多半表示状态的变化。其意义有两种:"使得……"或"变得……"。例如,Her face was *wet* with tears. (*adj.* 潮湿的) She walked carefully so as not to *wet* her beautiful shoes. (*v.* 使潮湿)

III. Read the following sentences and tell the Chinese meanings of the underlined words.
1. 1) There must be something wrong with my computer, for it is getting warmer and warmer.
   2) Martin's words warmed my heart.
2. 1) Usually we should put our computers in a dry place.
   2) Dry your clothes and don't let the water drop on the monitor.
3. 1) That space is too narrow. I cannot put my desktop computer there.
   2) Narrow the margin of this page, or you cannot type in the whole paragraph.

4. 1) My laptop is too <u>slow</u>. I cannot wait any more.
   2) <u>Slow</u> down your speed to avoid traffic disaster.

Across

1. Word-processing _____s typing, but one can modify any part of the text he typed in.
2. Connect VGA (Video Graphics Array) _____ on your computer to a TV set, and you can watch the movie in your computer on TV.
3. You should be very careful before you _____ your hard drive. Otherwise, you will lose all the information in it.
4. A computer is a _____ for processing information.
5. To be a computer _____, you need to use it very often.

Down

1. Even a simple database is a great help for us to _____ millions of information.
2. Putting different information into different _____s is an easy way for you to find them later.
3. One of the most popular _____s now is Microsoft Internet Explorer.
4. It is hard to _____ the text you typed on a typewriter.
5. The best way to _____ your idea is by using PowerPoint.

Unit 4 Basic Computer Skills

Fun Time

Customer: Your sound card is defective and I want a new one.
Tech Support: What seems to be the problem?
Customer: The balance is backwards. The left channel is coming out of the right speaker and the right channel is coming out of the left. It's defective.
Tech Support: You can solve the problem by moving the left speaker to the right side of the machine and vice versa.

Project

Communicate with others!

Step 1: Within each group, talk about how to create a folder and name it "Homework".

Step 2: Within each group, talk about how to create another folder and name it "English Homework", and move it into the folder of Homework.

Step 3: Ask one classmate from each group to give an oral presentation to tell the whole process in class.

Self-checklist

根据实际情况，从A、B、C、D中选择合适的答案：A代表你能很好地完成该任务；B代表你基本上可以完成该任务；C代表你完成该任务有困难；D代表你不能完成该任务。

A B C D

☐ ☐ ☐ ☐   1. 能掌握并能运用本单元所学重点句型、词汇和短语。

☐ ☐ ☐ ☐　2. 能理解并正确模仿听读部分的句子，正确掌握发音及语调。
☐ ☐ ☐ ☐　3. 能模仿句型进行简单的对话。
☐ ☐ ☐ ☐　4. 能读懂本课的短文，并正确回答相关问题。
☐ ☐ ☐ ☐　5. 能掌握不定式的用法。
☐ ☐ ☐ ☐　6. 能了解转化构词法中形容词与动词之间的相互转化。
☐ ☐ ☐ ☐　7. 能用课文中学习的词组造句。
☐ ☐ ☐ ☐　8. 能向同学介绍使用电脑所需的基本技能。

Unit 5

Word Processing Software

Unit Goals

In this unit, you will be able to
- understand what we can do with a word processor;
- talk about word processors;
- understand articles about word processors;
- write a short paragraph about word processors;
- demonstrate your knowledge about word processors.

Lead-in

Picture matching.

| title bar | text area | Ribbon |
| scroll bar | file button | toolbar |

1.
2.
3.
4.
5.
6. The word processing screen

Listening and Speaking

Listen and complete.

a. It's a great help to me.
b. Could you tell me what you are doing with the word processing software?
c. I'm creating a Christmas card.
d. Well, I use it to write articles for my magazine.

What Can We Do with a Word Processor?

Different people use a word processor to do different things. Jeanette, a girl from a senior high school, is asking people some questions concerning a word processor.

She is talking with Ms Lee, a journalist.

❶ Jeanette: Good morning, Ms Lee. What do you do with a word processor?

Ms Lee: _____

She is asking her brother, Mike.

❷ Jeanette: Hi, Mike. What do you do with a word processor?

Mike: I use it to do my assignments and write reports.

_____

She is asking Miss Zhang, a secretary in an office.

❸ Jeanette: Excuse me, Miss Zhang. _____

Miss Zhang: I'm writing a report. Our company is losing money and we are writing a report to the Board of Directors.

Unit 5 Word Processing Software

She is talking to her mom.

④ Jeanette: Hi, mom. What are you doing with a word processor?

Mom: _____ I am going to send it to each of our friends with our seasonal greetings. Aren't you going to make one of your own?

Role Play

A: Good morning, Ms / Mr / Miss ... What do you do with a word processor?

B: Well, I use it to write articles / do my homework / make Christmas cards ...

Words & Expressions

toolbar /'tu:lbɑ:(r)/ n. 工具栏
Ribbon /'rɪbən/ n. 功能区
title /'taɪtl/ n. 名称；标题
article /'ɑ:tɪkl/ n. 文章；论文
senior /'si:niə(r)/ adj. 年长的；高级的

journalist /'dʒɜ:nəlɪst/ n. 新闻记者
assignment /ə'saɪnmənt/ n. 作业
seasonal /'si:zənl/ adj. 季节的

Board of Directors 董事会

Reading

Text A

Pre-reading activities.

1. Do you know these icons?

2. Where can we find these icons? Discuss their functions with your partner.

3. Can you guess what the text talks about when you see these icons and the title of the text?

Reading Strategy

预测 (Predicting) 预测策略就是根据文章的标题或图示等预测文章的内容。也可根据段落中的第一个句子来推测本段的内容，或是根据已读段落的内容来推测未读段落的内容。

How to Use Word Processors?

Word processors allow users to enter, edit and save text. Users can also print the text with the help of a printer.

Hitting the keys on the keyboard, users will enter text. You can see whatever you type on the screen. Text can be added or deleted at any point, letter-by-letter or in blocks.

Text can be shown in many styles, such as in bold, italic or underlined. You can also change its font, size and colour. Text can be moved from place to place in a document or between documents.

You need to save your document often in case it gets lost. To save the text, you can go to the "File" menu, then find and click "Save As" button. Remember to give your text a name and locate it at a proper place so that you can easily find it later. You can also print the text by pressing "Print" button on the toolbar.

Words & Expressions

edit /ˈedɪt/ v. 编辑，校订
　　　　n. 编辑工作
delete /dɪˈliːt/ v. 删除
block /blɒk/ n. (一) 批
bold /bəʊld/ adj. 粗体的
　　　　n. 黑体字，粗体字
italic /ɪˈtælɪk/ adj. 斜体的
　　　　n. 斜体

underline /ˌʌndəˈlaɪn/ v. 在……下面画线
　　　　　　n. 下画线
font /fɒnt/ n. 字体，字形
document /ˈdɒkjʊmənt/ n. 文件
locate /ləʊˈkeɪt/ v. 定位
proper /ˈprɒpə(r)/ adj. 合适的

in case 以免，万一

True or false.
1. A word processor is used to draw pictures.
2. Text can be both added and deleted.
3. We can move a text from one document to another.
4. Documents should be saved every now and then.
5. Each document should have its own name.

Text B

Pre-reading questions.
1. Do you know why word processors are widely used?
2. Please list some basic features of word processors.

Why Do We Use Word Processors?

A word processor is a computer application used to produce any kind of printable material.

Now the computer is used widely for word processing. It provides an easy way to create letters, reports, résumés and other printed documents. In order to fully understand its applications, you must get rid of the idea that word processing is just typing words on a page. It can provide countless easy-to-use and time-saving features. Word processors have become indispensable for the office work today.

Word processors are quite different from each other, but all of them support the following basic features: insert text, delete text, set page size and margins, search and replace, print and so on. There are many other unexpected features of word processors that are great, but these should be enough to get you started.

Words & Expressions

printable /ˈprɪntəbl/ *adj.* 可打印的
résumé /ˈrezjʊmeɪ/ *n.* 简历
countless /ˈkaʊntləs/ *adj.* 无数的
time-saving /ˈtaɪm seɪvɪŋ/ *adj.* 省时的
indispensable /ˌɪndɪˈspensəbl/ *adj.* 不可缺少的
insert /ɪnˈsɜːt/ *v.* 插入

margin /ˈmɑːdʒɪn/ *n.* 页边的空白
replace /rɪˈpleɪs/ *v.* 替换
unexpected /ˌʌnɪkˈspektɪd/ *adj.* 意想不到的

in order to 为了
get rid of 除掉
be different from 与……不同

Unit 5 Word Processing Software

Complete the passage with the words in the box.

| delete | insert | text | document | pointer | printer |

The great advantage of word processing is that you can make changes without retyping the entire _____. If you make a typing mistake, you simply back up the _____ and correct your mistake. If you want to _____ a paragraph, you simply remove it, without leaving a trace (痕迹). It is equally easy to _____ a word, sentence, or paragraph in the middle of a document. Word processors also make it easy to move sections of _____ from one place to another within a document, or between documents. When you have made all the changes you want, you can send the file to a _____ to get a hard copy.

Notes

使用鼠标和键盘可以迅速地选出需要改变的部分文本。

目的 (Purpose)	操作 (Action)
选择一个单词	双击该单词
选择几个单词	点击第一个单词，拖动至最后一个单词
选择一个句子	按下CTRL键并点击句子
选择一行	点击左边空白
选择一段	在段落内三击鼠标或双击左边空白
选择整个文本	CTRL+A或三击左边空白

Do You Know

汉字输入法

　　汉字输入法，是指为了将汉字输入计算机或手机等电子设备而采用的编码方法，是中文信息处理的重要技术。输入法分为两大类：键盘输入法和非键盘输入法。键盘输入法编码可分为几类：音码、形码、音形码、形音码、无理码等。广泛使用的中文输入法有拼音输入法、五笔字型输入法、二笔输入法、郑码输入法等。流行的输入法软件平台，在Windows系统有搜狗拼音输入法、搜狗五笔输入法、百度输入法、谷歌拼音输入法、QQ拼音输入法、QQ五笔输入法、极点中文汉字输入平台；手机系统一般内置中文输入法，此外还有百度手机输入法、搜狗手机输入法等。

　　非键盘输入法包括字形识别输入法和语音识别输入法。前者，如各种形式的手写板，用笔在一块特制的板上书写，即可把所写文字输入计算机。后者通过语音识别技术，能把人的语音转换为计算机可以接受的数字信号，并把汉字显示在屏幕上。

快捷键使用技巧

[CTRL] + X	剪切	[CTRL] + C	复制
[CTRL] + V	粘贴	[CTRL] + F	查找
[CTRL] + H	替换	[CTRL] + A	全选
[CTRL] + S	保存		

 Language Practice

I. Fill in the blanks with the missing letters of the words according to the pictures.

1. wr_te rep_ _ts

2. a Chri_ _m_s c_ _d

3. a j_ _ _nal_st

4. a m_g_zine

Unit 5 Word Processing Software

5. some p_r_graphs

6. gr_ _ t

7. it_lic w_ _ds

8. h_ndwrit_ _ _

9. m_n_

I am happy today!

10. _nd_ _line

II. Complete the sentences with the words practiced above.

1. Jude's _____ is quite good.
2. Peter _____ for several newspapers.
3. The _____ are the most important.
4. Last Christmas, Ann received many _____.
5. If you want to print your document, you can find "print" in the "file" _____.
6. As a _____, Mr Smith has been to 15 countries.
7. Please buy a _____ for me on your way home.
8. John, read this _____, please.
9. When you meet your classmates in the morning, you may _____ them with "Good morning".
10. _____ the words you want to emphasise.

III. Complete the passage with the words in the box.

> text area   status bar   toolbar   Ribbon   ruler   title bar

Most word processing windows contain the following elements:
The _____ has the name of the word processor.
The _____ has a list of pull-down menus.
The _____ displays a set of buttons to carry out common commands.
There are some numbers on the _____.
The _____ is where the text will be displayed.
The _____ shows the page and line numbers.

IV. Choose the proper word to complete the passage.

I had been doing Tech Support for Hewlett-Packard's DeskJet division for about a month _____ (when/while) I had a customer call with a problem I just couldn't solve. She could not print yellow. All the _____ (another/other) colours would print fine, which truly confused me because the only true colours are cyan (蓝色), magenta (红色), and yellow. I had the customer _____ (change/changed) ink cartridges (墨盒), delete and reinstall the drivers. Nothing worked. I asked my coworkers _____ (for/about) help; they _____ (afforded/offered) no new ideas. After over two hours of troubleshooting, I was about to tell the customer to send the printer in to us for repair when she asked _____ (quietly/quite), "Should I try printing _____ (on/in) a piece of white paper instead of yellow paper?"

V. Answer the questions.

How to Print a Document?

There are two ways to have your document printed.

Tip 1

Click the print icon on the toolbar to print quickly, but you cannot change the options in this way.

Tip 2

You can hit "Ctrl" + "P" on your keyboard, or go to the "File" menu to access the printing options. There are several options, such as the number of copies to print, the exact pages to print, which printer to use and so on.

List two ways to print a document (with less than 8 words in each answer).

Tip 1: _____

Tip 2: _____

VI. Match the meanings.

1. The spell-checker	a. can make your document clear and easy to understand.
2. Touch-typing	b. can find out spelling or typing mistakes.
3. Paragraphs, titles and bullets	c. can help you type faster.

VII. Sentence completion.

1. You need to save your document often _____ it gets lost. (你需要经常保存文件以防丢失。)

2. Martin bought a word processor application handbook _____ fully understand it. (为了更好地了解文字处理系统的应用，马丁买了一本操作手册。)

3. You must _____ the idea that word processing is just typing words. (你必须抛弃文字处理系统仅仅可以用来打字的这一想法。)

4. Word processors _____ from each other. (文字处理系统之间有很大的不同。)

5. Remember to locate the text at a proper place _____ you can easily find it later. (记住把文本存放在合适的路径下，以便将来能容易地找到。)

Writing

Fill in the blanks with the information you have got from the texts.

A word processor is a computer _____ used to produce any kind of _____. It provides an easy way to create letters, reports, résumés and other printed _____. It also can provide _____ easy-to-use and time-saving _____. Word processors have become indispensable for the _____ today. Word processors are quite _____ from each other, but all of them support the following basic features: insert text, _____ text, set page size and _____, search and replace, print and so on.

Grammar

	Present Participle (现在分词)
predicative	You must get rid of the idea that word processing is just **typing** words on a page.
attribute	Word processors are quite different from each other, but all of them support the **following** basic features.

Unit 5 Word Processing Software

（续表）

adverbial	of time	**Turning** around, Jimmy saw a car rushing to him.
	of cause	**Being** ill, Julia cannot come to the IT class.
	of condition	**Hitting** the keys on the keyboard, users will enter text.
	of concession	**Weighing** less than 1 kilogram, the new laptop is very popular among girls.
	of result	The storm in the northern part of Russia went on for several days, **killing** five people.
	of manner	**Singing** and **laughing**, the students went into the classroom.
complement	of object	I saw Jason **entering** the computer lab.
	of subject	The water must be kept **boiling** for another ten minutes.

I. Turn the underlined part of each sentence into an -ing phrase.
1. <u>As the secretary was away</u>, Mr Smith had to type his document.
2. John played computer games all day <u>and had no time left to do his homework</u>.
3. <u>When the computer class was over</u>, the students went to the playground.
4. <u>If time permits</u>, I am going to finish this typing work.
5. Joseph sat in a chair <u>and played computer games</u>.

II. Fill in the blanks with the correct forms of the verbs in the brackets.
1. Lily went to the store _____ (buy) a new computer yesterday.
2. There's no use _____ (argue) any more with him about the benefits of using a laptop.
3. I use a word processor _____ (make) the typing easier.

4. Joe spends all his leisure time _____ (play) computer games.
5. Do not forget _____ (back up) the data.

Word formation — Derivation (派生法)

派生词是由词根(主要是自由词根)加派生词缀构成。词根是派生词的基础，同一词根加不同词缀可表示不同的意义或不同的词性，大多数前缀并不影响词根的词性，而仅对词根的意义加以修饰，表示否定或相反的意思等，如：un-; in-; im-; il-; ir-; dis-; de-; non-; counter- 等。

Adjective / Verb / Noun	Prefix	Adjective / Verb / Noun
possible	im-	impossible
polite	im-	impolite
dependent	in-	independent
fair	un-	unfair
happy	un-	unhappy
do	un-	undo
agree	dis-	disagree
smoker	non-	non-smoker

III. Complete the sentences with the derivatives above.

1. It is _____ that not everyone got the chance to vote.
2. One mistake can _____ all our achievements.
3. It is _____ for the fish to live without water.
4. I _____ with what you have said.
5. The equipment has its own _____ power supply.
6. Catherine lives an _____ life in the country.
7. Nobody will be allowed to smoke in public places if there are _____ present.
8. It is _____ to say rude words to others.

Unit 5 Word Processing Software

Game

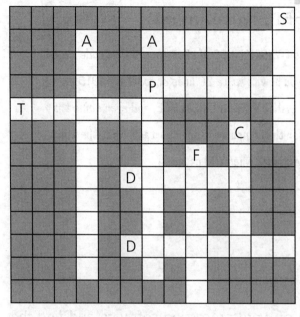

Across

1. In his middle twenties he began to write _____s and short stories.
2. You can send the file to a _____ to get a hard copy of the document.
3. Generally there are three buttons, "save", "withdraw", "retrieve" in the _____ of word processors.
4. Text can be added or _____d at any point in word processors.
5. Secretaries use computers to manage their _____s.

Down

1. The _____ of word processors is popular today.
2. Journalists use a word _____ to write articles for magazines.
3. Insert text, delete text and print are the basic _____s of word processors.
4. A word processor provides an easy way to _____ letters, reports and other printed documents.
5. One can direct the word processor to _____ for a word or phrase very easily.

Fun Time

A New Kind of Programming

They say that a GOOD programmer can write TWENTY LINES of effective program code a day! With our unique training system, we'll show you how to write 20 lines of code, and LOTS more! Our course covers EVERY PROGRAMMING LANGUAGE IN EXISTENCE, and even some that aren't! You'll learn why the ON / OFF switch for the computer is so important, what the words "FATAL ERROR" mean, and even who should be blamed when YOU cause it!

Project

Communicate with others!

Step 1: Try to get an English version of word processing software and find the file buttons.

Step 2: Within each group discuss the functions of each file button.

Step 3: Ask one classmate from each group to report the result of their discussion.

Self-checklist

根据实际情况，从A、B、C、D中选择合适的答案：A代表你能很好地完成该任务；B代表你基本上可以完成该任务；C代表你完成该任务有困难；D代表你不能完成该任务。

A B C D

☐ ☐ ☐ ☐ 1. 能掌握并能运用本单元所学重点句型、词汇和短语。

☐ ☐ ☐ ☐ 2. 能理解并正确模仿听说部分的句子，正确掌握发音及语调。

Unit 5 Word Processing Software

☐ ☐ ☐ ☐   3. 能模仿句型进行简单的对话。
☐ ☐ ☐ ☐   4. 能读懂本课的短文，并正确回答相关问题。
☐ ☐ ☐ ☐   5. 能掌握现在分词的用法。
☐ ☐ ☐ ☐   6. 能掌握以 un-; in-; im-; il-; ir-; dis-; de-; non-; counter- 为前缀的派生词构词方法。
☐ ☐ ☐ ☐   7. 能用课文中学习的词组造句。
☐ ☐ ☐ ☐   8. 能向同学介绍文字处理系统的基本知识。

Unit 6

Spreadsheets

Unit Goals

In this unit, you will be able to

- understand listening materials about what we can do with a spreadsheet;
- talk about spreadsheets;
- understand articles about basic spreadsheet knowledge;
- write a short paragraph about spreadsheets;
- demonstrate your knowledge about what a spreadsheet is.

Lead-in

Picture matching.

column	formula	cell	function	rows
C3	label	value	scroll button	

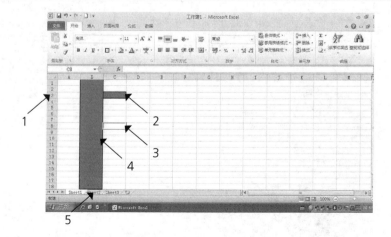

Unit 6 Spreadsheets

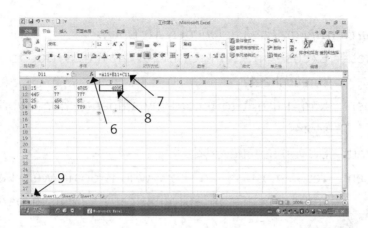

Listening and Speaking

Listen and complete.

- a. It's a great help to me.
- b. I'm trying to lose weight.
- c. Well, I run a local tennis league.
- d. It tells what I have sold.

What Can We Do with a Spreadsheet?

Brett is a salesman of a computer store. He is doing a market survey to see how people use the spreadsheet software in their daily life.

He is now greeting Mr Granger, a regular customer and owner of a store around the corner.

 Brett: Good morning, Mr Granger.
Mr Granger: Morning, Brett.
Brett: What do you think of the spreadsheet software?
Mr Granger: Terrific. I use it to manage my store. _____ In this way, it can help me keep good accounts.

83

He is interviewing an English teacher, Mr Chen.

2 Brett: Excuse me, Mr Chen. What do you do with the spreadsheet?

Mr Chen: Mm, I use it to record my students' homework and test scores.

Brett: Do you use it everyday?

Mr Chen: Yes, almost. It can show who is doing well. Besides, the spreadsheet can automatically calculate percentages.

_____

He is now with his friend, Linda.

3 Brett: Hi, Linda. Haven't seen you for a long time. What are you doing with the spreadsheet?

Linda: _____ So I use a spreadsheet to keep track of my eating habits.

Brett: Do you keep a record of your diet?

Linda: Yeah. In this way I can know how many calories I've taken. A spreadsheet can also tell me if I'm getting enough vitamins.

He is talking to Mr Brooks, a coach of a tennis league.

4 Brett: Excuse me, Mr Brooks. Could you spare me a few minutes and tell me what you do with a spreadsheet?

Mr Brooks: _____ And I use a spreadsheet often.

Brett: Is it very helpful?

Mr Brooks: Yeah. I use a table of results and create a ranking for each player.

Unit 6 Spreadsheets

Role Play

A: Excuse me, Ms / Mr / Miss ... What do you do with a spreadsheet?

B: Well, I use it to manage my store / record my students' homework and test scores / keep track of my eating habits / keep a table of results for each player ...

Words & Expressions

column /ˈkɒləm/ *n.* 圆柱；列
cell /sel/ *n.* 单元格
label /ˈleɪbl/ *n.* 标签；商标
function /ˈfʌŋkʃn/ *n.* 函数
formula /ˈfɔːmjələ/ *n.* (pl. formulas 或 formulae) 公式；规则
graph /ɡrɑːf/ *n.* 图表；曲线图
value /ˈvæljuː/ *n.* [数]值
survey /ˈsɜːveɪ/ *n.* 民意调查
regular /ˈreɡjələ(r)/ *adj.* 经常的，频繁的
terrific /təˈrɪfɪk/ *adj.* (口) 好极了

account /əˈkaʊnt/ *n.* 账目
interview /ˈɪntəvjuː/ *v.* 采访
score /skɔː(r)/ *n.* 得分
automatically /ˌɔːtəˈmætɪkli/ *adv.* 自动地
calculate /ˈkælkjuleɪt/ *v.* 计算
calorie /ˈkæləri/ *n.* 卡路里 (热量单位)
ranking /ˈræŋkɪŋ/ *n.* 顺序；等级

keep track of 了解……的动态
keep a record of 记录

Reading

Text A

Pre-reading activities.

1. Guess the meanings of the following words with your knowledge of word formation.

| recalculate | numerical | limitless |
| highlight | update | addition |

85

2. Do you know what Excel is?

3. What are the advantages of Excel?

Reading Strategy

词义猜测 Word Guessing 在阅读中遇到生词时不要急于查字典，而是分析生词本身的构词特点，利用构词法的相关知识，如复合、派生等方法来推断词义。

The Advantages of Excel

Excel is a spreadsheet program. People may use it to enter numerical values or data into the rows or columns of a spreadsheet, and to use them for such things as calculations, graphs, and statistical analysis.

A spreadsheet has many advantages. Users can "program" it to perform certain functions automatically, such as addition, subtraction, etc. Although you can do simple calculations in Word, it is much easier to use Excel. Each figure will be input in its own cell and it is so easy to use formulas to add up, subtract, divide or multiply. For example, you have 1 000 numbers to add up. With Excel you just highlight them all and click on the sum icon and they add up immediately. Change a few numbers and the total will recalculate. The result will always be correct.

A spreadsheet also has the ability to hold almost limitless amounts of data. Once created, the spreadsheet can be quickly updated.

Unit 6 Spreadsheets

Words & Expressions

numerical /njʊ(ː)'merɪkl/ *adj.* 数字的，用数表示的
statistical /stə'tɪstɪkl/ *adj.* 统计的
analysis /ə'næləsɪs/ *n.* 分析
perform /pə'fɔːm/ *v.* 运行
subtraction /səb'trækʃn/ *n.* 减去，消减
divide /dɪ'vaɪd/ *v.* [数] 除
multiply /'mʌltɪplaɪ/ *v.* [数] 乘

highlight /'haɪlaɪt/ *v.* 选中；使显著
immediately /ɪ'miːdiətli/ *adv.* 立即，马上
limitless /'lɪmɪtləs/ *adj.* 无限的
update /ˌʌp'deɪt/ *v.* 更新

add up 加起来

Short-answer questions.

Answer the following questions according to Text A.

1. What can we do with the help of Excel?

   _____

2. Why do we use Excel to do calculations instead of Word?

   _____

3. Can Excel do addition automatically under a certain condition?

   _____

4. Do we need add the figures one by one in Excel?

   _____

5. How much data does a spreadsheet hold at most?

   _____

87

Text B

Pre-reading questions.
1. What is your ID (Identification) number?
2. What do you usually do when you have trouble with computer software?

An ID Ten T Error

Judy, a new clerk in our office, was having trouble with her spreadsheet software. So she called Peter, who always had a good sense of humour, over to her desk. He clicked several buttons and solved the problem.

As Peter was walking away, Judy called after him, "So, what was wrong with my computer?" Peter answered, "It was an ID ten T error."

"What does that mean?" Judy asked. He gave her a smile and asked, "Haven't you ever heard of an ID ten T error before? Write it on a piece of paper, and you will see."

Judy wrote "ID10T" (IDIOT) on a sheet of paper. Then all of a sudden she cried out "idiot". They looked at each other and burst into laughter.

Words & Expressions

identification (ID) /aɪˌdentɪfɪˈkeɪʃn/ *n.* 身份
error /ˈerə(r)/ *n.* 错误
clerk /klɑːk/ *n.* 职员
sense /sens/ *n.* 感觉
humour /ˈhjuːmə(r)/ *n.* 幽默
solve /sɒlv/ *v.* 解决
idiot /ˈɪdɪət/ *n.* 傻瓜
sheet /ʃiːt/ *n.* (一) 张

sudden /ˈsʌdn/ *n.* 突然发生的事 (只用于习语)
adj. 突然的
burst /bɜːst/ *v.* 爆发
laughter /ˈlɑːftə(r)/ *n.* 笑；笑声

hear of 听说
all of a sudden 忽然，一下子

Unit 6 Spreadsheets

True or false.
1. Peter is a serious guy.
2. It was not difficult for Peter to solve the problem.
3. Peter taught Judy how to solve the problem.
4. Judy could not understand the meaning of "ID ten T" at first.
5. "ID ten T (idiot)" means a foolish person.

Notes

cell 单元格：Excel工作表的基本元素，其中的小格均称为单元格，能存放输入的数据和文字。单元格在工作表中的位置称为单元格地址，用列坐标和行坐标表示。列坐标用英文字母表示，共256列；行坐标用自然数表示，共65 536行。所以，一张工作表最多有65 536 × 256个单元格。

function 函数：是公式的一种。在Excel中，函数是预先定义好的公式，如数值求和函数、求最大值函数、求平均值函数等。

Do You Know

Excel 电子表格软件

　　Excel是Office系列办公软件中的一个电子表格软件，用于制作电子表格、完成复杂的运算、进行数据分析及预测、制作图表。它具有界面友好、操作简单、易学易懂等特点，并且引入了公式和函数计算、统计图表绘制和图形绘制等功能，能有效管理、分析数据和制作网页等。

Excel 电子表格中的几个常用函数
　　SUM: 计算多个参数的总和。
　　AVERAGE: 计算多个参数的平均值。

> IF: 判断一个条件是否满足，如果满足返回一个值，如果不满足返回另一个值。
> COUNT: 计算包含数字的单元格以及参数列表中的数字的个数。
> SUMIF: 对满足条件的单元格求和。
> MAX: 返回一组数值中的最大值，忽略逻辑值及文本。

Language Practice

I. Fill in the blanks with the missing letters of the words according to the pictures.

1. c_ll

2. f_ _mul_

3. an _dent_fic_tion card

4. a s_lesm_n

5. t_nn_s

6. being _nt_ _v_ _wed

8. sc_ _e

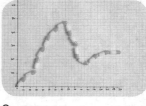

9. gra_ _

10. _cc_ _nts

7. a cl_ _k

Unit 6 Spreadsheets

II. Complete the sentences with the words practiced above.

1. The first duty of an accountant is to keep clear _____.
2. The young man is _____ by a reporter, for he saved a boy yesterday.
3. Xiao Yang, one of my former classmates, is a bank _____ now.
4. Put these numbers in the middle of the _____.
5. As a _____, you must be good at persuading others into buying your products.
6. The _____ in the basketball game was 112:107.
7. With that _____, you may work out the problem easily.
8. Paul and his classmates made a weather _____ of this month.
9. Would you like to play _____ with me this Sunday?
10. Show me your _____ card, please.

III. Fill in the blanks with the words in the box.

formula	label	function	value

How to Enter Data into a Spreadsheet

1. You can enter texts as _____, such as people's names or titles of CDs.
2. You can also enter numbers like dates or money as _____.
3. A _____ is used to process some data, such as adding up all the numbers in a column and then show the result in a cell at the bottom of the column.
4. A _____ is a built-in formula. It is a shortcut for common calculations such as subtraction and division.

Ⅳ. Here are some tips for spreadsheets.

Tip 1

Fill in the blanks with the verbs given below according to the given picture.

| type | add | select |

How to Write Formulas

_____ the cell you wish to have the result in.

Type "=".

Then _____ the function you wish to perform (运行), using cell names as subjects. For example, "= E2 + F2" would _____ the value in E2 to the value in F2.

Tip 2

 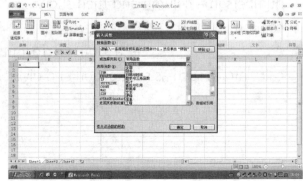

Unit 6 Spreadsheets

Change the order of the following three steps.
How to Use Functions

1. _____
2. _____
3. _____

| a. Click the function. | b. Choose a category from the drop down menu. | c. Go to the "Function", click it. |

Tip 3
Fill in the blanks with the words in the box.

| spreadsheet | help | use | tables |

Once information has been entered into a _____, it can be calculated easily. We can produce _____ of results, graphs and make calculations based on our results. We can also _____ "+" for addition, "-" for subtraction, "*" for multiplication and "/" for division in our formula.
Use the "_____", if you need to, by clicking or pressing "F1".

V. Match the types of the charts with the words in the box.
People use charts in their daily work and study. The followings are some commonly used charts.

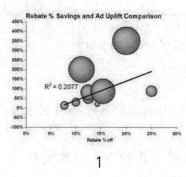

1

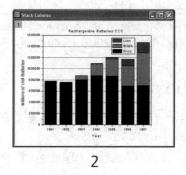

2

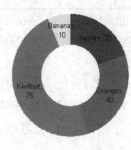

3

93

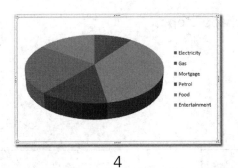
4

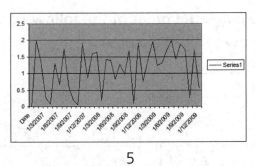
5

| Doughnut | Pie | Line | Bubble | Column Bar |

VI. Write down the formulas that would be entered for each cell (1), (2), (3) and (4).

	A	B	C	D	E	F	G	H	I
1	No.	Name		Oral Score		Test Score		Final Score	
2	1	Sally		38.4		62.5		69.7	
3	2	Allen		42.95		76.5		81.2	
4	3	Helen		31.1		57.5		(1)	
5	4	Sarah		38.85		70.5		74.1	
6	5	Alice		41.55		82.5		82.8	
7	6	Garry		39.65		77.5		(2)	
8	7	Lee		37.95		56.5		66.2	
9									
10		>=90	80-89	70-79	60-69	<60			Average
11		0	2	(4)	2	1			(3)

(The full mark of the oral score is 50, which accounts for 50% of the final score;
The full mark of the test score is 100, which also accounts for 50% of the final score.)

VII. Sentence completion.

1. I use a spreadsheet to _____ my eating habits. (我用电子数据表来了解自己饮食习惯的变化。)

2. Do you _____ how much you spend? (你对自己的开支记账吗?)

3. _____ John cried out. (约翰突然哭了起来。)
4. Scott has a good _____. (斯科特很有幽默感。)
5. Please _____ the column of figures. (请把这栏数字加起来。)

Writing

Fill in the blanks with the information you have got from the texts.

Excel is a _____ program. People may use it to enter numerical values or data into the _____ or _____ of a spreadsheet, and thus use them for such things as _____, graphs, and statistical analysis. Although you can do simple calculations in _____, it is much easier to use Excel. In Excel, change a few numbers and the total will _____. The result will always be correct. Once created, the spreadsheet can be quickly _____.

Grammar

Past Participle (过去分词)		
predicative		Once created, the spreadsheet can be quickly **updated**.
attribute		Julia is the only one **left** after the class.
adverbial	of time	**Seen** from a distance, the hill you drew on your computer looks like a rabbit.
	of cause	Deeply **hurt**, Allen ran out of the room.
	of condition	Once **created**, the spreadsheet can be quickly updated.
	of concession	Although **given** the best medical care, Mark didn't survive.

（续表）

	of result	Carl worked so hard and achieved a lot in the genetic field, thus **awarded** Nobel Prize.
	of manner	Alex and his family arrived home at last **depressed**.
complement	of object	I found the cleaning work completely **finished**.
	of subject	The cleaning work was found completely **finished**.

I. Multiple choices.

1. What is the formula _____ in this calculation?
   A. using　　　　　　B. used　　　　　　C. to use
2. Andy had his test scores _____ with a spreadsheet.
   A. to record　　　　B. recorded　　　　C. record
3. Adam is upstairs _____ on Joe's computer.
   A. works　　　　　B. is working　　　　C. working
4. _____, the students tried their best to do the calculation.
   A. Greatly encouraged
   B. To be greatly encouraged
   C. Greatly encouraging
5. _____, the subject belongs to computer science.
   A. Strictly spoken　　B. Strictly speaking　　C. Strictly speak

II. Fill in the blanks with the correct forms of the verbs in the brackets.

1. Never close the editor before the file is _____ (save).
2. Brooks wants his computer _____ (upgrade).
3. _____ (update), the software works well.
4. Aidan found the network _____ (break).
5. The computer virus is reported _____ (control).

Word formation — Derivation (派生法)

在派生构词法中，一般后缀会改变词性，前缀改变词义，但也有些后缀既改变了词性又改变了词义。形容词加-ly变成副词；名词加-less变成形容词，且变成相反的词义。

Adjective	Suffix	Adverb
immediate	-ly	immediately
high	-ly	highly
clear	-ly	clearly
careful	-ly	carefully

Noun	Suffix	Adjective
limit	-less	limitless
care	-less	careless
home	-less	homeless
hope	-less	hopeless

III. Complete the sentences with the words above.

1. Sally went back to her seat _____ after she made her presentation.
2. We gazed out over the _____ expanse of the desert.
3. You are too _____. How can you put a glass of water so close to your computer.
4. Please type the document _____.
5. We can get some information about _____ people via the Internet.
6. Diana is a _____ educated and skillful programmer.
7. Tony became _____, since his computer broke twice a day.
8. This book explains the application of Word very _____.

Game

Across

1. With Excel, it is easy to use _____s to add up, subtract, divide and multiply.
2. People may use Excel to enter data into the rows or _____s of a spreadsheet.
3. Yao Ming is being _____ed by some reporters.
4. Please _____ the words you want to emphasise.
5. Excel is a _____ program.
6. Tony _____ into laughter for no reason.

Down

1. The new _____ in our office was having trouble with her spreadsheet software.
2. I use a table of results and create a _____ for each player.
3. The database _____s automatically when new information is entered.
4. A spreadsheet has the ability to hold almost _____ amounts of data.

Fun Time

Computer Stupidities

Last year, the temp agency I was working for was arranging a contract for me, and some

Unit 6 Spreadsheets

additional "computer skills" tests were necessary. The branch manager asked what kind of computer I was comfortable with. I said, "Windows PC," although I had used several others. She cut in right then and asked, "Word or Excel?"

Project

Communicate with others!

Step 1: Try to collect the information on the age and the weight of your classmates and record the data with the spreadsheet.

Step 2: Within each group, talk about what formulas you use to calculate their average age and average weight.

Step 3: Within each group, talk about what formulas you use to calculate the number of students below the average and above the average.

Self-checklist

根据实际情况，从A、B、C、D中选择合适的答案：A代表你能很好地完成该任务；B代表你基本上可以完成该任务；C代表你完成该任务有困难；D代表你不能完成该任务。

A B C D

☐ ☐ ☐ ☐ 1. 能掌握并能运用本单元所学重点句型、词汇和短语。

☐ ☐ ☐ ☐ 2. 能理解并正确模仿听说部分的句子，正确掌握发音及语调。

☐ ☐ ☐ ☐ 3. 能模仿句型进行简单的对话。

☐ ☐ ☐ ☐ 4. 能读懂本课的短文，并正确回答相关问题。

☐ ☐ ☐ ☐ 5. 能掌握过去分词的用法。

☐ ☐ ☐ ☐ 6. 能掌握以-ly, -less为后缀的派生词的构词方法。

☐ ☐ ☐ ☐ 7. 能用课文中学习的词组造句。

☐ ☐ ☐ ☐ 8. 能向同学介绍电子数据表的基本知识。

Unit 7 Database

Unit Goals

In this unit, you will be able to
- understand listening materials about the functions of databases;
- talk about databases;
- understand articles about basic database knowledge;
- write a short paragraph about databases;
- demonstrate your database knowledge.

Lead-in

Picture matching.

database file   field   record   table   data type

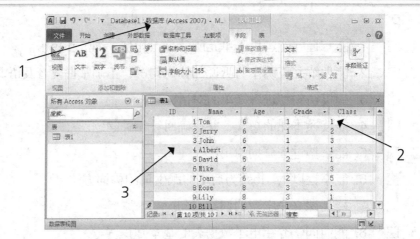

Unit 7 Database

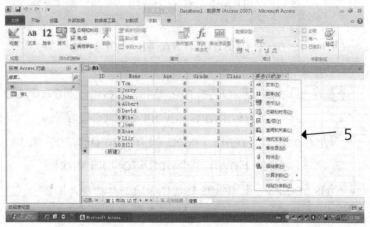

Listening and Speaking

Listen and complete.

a. It keeps records of all the staff information so that I can search and retrieve any information I need.
b. The records in the database tell us the history of each patient.
c. How is it useful for your corner shop?
d. It helps us see how many rooms are available.

Do You Use Databases?

Different people use databases for different purposes. Dave is asking people what they do with databases.

101

He is talking with Mr Brooks, a grocery store owner.

1   Dave: Good morning, Mr Brooks. Do you have database software installed in your computer?

Mr Brooks: Of course. And it's a great help to me.

Dave: _____

Mr Brooks: I use a database system to keep an inventory of my stores. In this way I can see what sells well, how much I earn each day and how much stock I have got.

Dave: Wow, your store is as organised as a big supermarket. Thank you.

Dave is now talking with Ms Spears, a boss of an Import & Export company.

2   Dave: Excuse me, Ms Spears. Have you installed a database program in your computer?

Ms Spears: Yeah. I keep all my employees' data in a database.

Dave: How is it helpful to your work?

Ms Spears: It helps the management a lot. _____

Dave: Thanks a lot.

He is talking with Monica, a receptionist in a hotel.

3   Dave: Good morning, Miss. Do you use databases in your work?

Monica: Yes. It's an important part of my work. Every time a customer checks in or out of our hotel, the information is entered into the database.

Dave: Sounds interesting. Is it helpful?

Monica: Definitely. _____ It never makes any mistakes.

Dave: Many thanks and good luck.

He is asking Mrs Bush, a nurse at Dental Care Clinic.

4   Dave: Good afternoon, Madame. Do you use databases at your clinic?

Mrs Bush: Yes. We have set up a database of all our patients.
Dave: Do you find it useful?
Mrs Bush: Yes, definitely. We can't imagine how we can manage without it. _____ Above all, the database has saved us a lot of room of storing those paper files.
Dave: Many thanks.

Role Play

A: Do you use database software in your work / study?
B: Yes, a lot.
   No, I don't use it at all.
A: What do you use a database for?
B: Well, I use it to keep records of my patients / students / ...

Words & Expressions

field /fi:ld/ n. 字段；信息组；栏
available /ə'veɪləbl/ adj. 可用的
inventory /'ɪnvəntri/ n. 详细目录；库存
retrieve /rɪ'tri:v/ v. 检索；重新获得
grocery /'ɡrəʊsəri/ n. <美>食品杂货店；食品，杂货
import /'ɪmpɔ:t/ n. 输入，进口
export /'ekspɔ:t/ n. 输出，出口
program /'prəʊɡræm/ n. 程序；节目

receptionist /rɪ'sepʃənɪst/ n. 接待员
definitely /'defɪnətli/ adv. 肯定；当然
dental /'dentl/ adj. 牙齿的
clinic /'klɪnɪk/ n. 诊所

set up 建立，树立
check in 签到；(旅馆、机场等)登记
check out 付账离开，结账

103

Reading

Text A

Pre-reading activities.

1. What is a relational database?

2. Try to guess the meaning of SQL Server and Open Source from the context.

Reading Strategy

词义猜测  除了利用构词法推测生词的意思外，通过分析生词与前后词的搭配、语法、上下文的内在关系等，也可以推测生词的意思，以使阅读顺畅。常见的引出同义词的标志性词语有：or、like、that is等；引出反义词的标志性词语有：yet、but、instead of 等。

Relational Database

A database is a computer application that manages data and allows fast storage and retrieval of data. There are different types of databases but the most popular is a relational database. It stores data in tables where each row in it holds the same sort of information.

In the early 1970s, Ted Codd, an IBM researcher, devised 12 laws of normalisation. These apply to the storage of data and relations between different tables. SQL is a simple programming language that is used in relational databases. SQL Server, a relational database developed and sold by Microsoft, and Oracle are popular commercial relational databases. Yet, open source relational databases also exist, such as MySQL. Open source means applications that come with full source code that you can use. There is an amazing amount of open source software on the Web and it makes an excellent way to learn by studying what others have written.

Words & Expressions

retrieval /rɪ'triːvl/ *n.* 数据检索
relational /rɪ'leɪʃənl/ *adj.* 关系的
devise /dɪ'vaɪz/ *v.* 设计
normalisation /ˌnɔːməlaɪ'zeɪʃn/ *n.* 规范化
commercial /kə'mɜːʃl/ *adj.* 商业的
amazing /ə'meɪzɪŋ/ *adj.* 令人惊异的

SQL 结构化查询语言
MySQL 一种开源关系型数据库
SQL Server SQL 服务器
open source 开放资源
source code 源代码

Short-answer questions.

Answer the following questions according to Text A.

1. What is a database?

2. What are the most popular databases?

3. Who devised the laws of normalisation for databases?

4. What is the language used in relational database?

5. Can we get full source code from open source relational databases?

_____

Text B

Pre-reading questions.
1. List some ways that people use to store information.
2. What is your usual way of storing data?

Ways of Storing Data

In the past the way of storing data was a card filing system. Card records can become files of papers that take up a large amount of space. It is often very difficult to find the information you need when the manual system becomes larger and larger.

Name: George Chan
Gender: Male
Age: 15
Nationality: Chinese
City: Beijing
Address: No. 6, Heping Road
Telephone: 010-2233 4455
School: Computer School

Now many people use computers to store data. Computers can store a large amount of information in a very small space. All the data are linked in a well-structured way, and can be retrieved easily and quickly.

When you buy database software, it has no files or data. The first thing you need to do is to set up a data file with collected data. It is also common to state what type of data will be entered into each field, and each field has its own name. Data can be in the form of numbers, characters, dates and so on.

Words & Expressions

manual /ˈmænjʊəl/ *adj.* 手动的，手工的
male /meɪl/ *n.* 男性
　　adj. 男的；雄性的
nationality /ˌnæʃəˈnæləti/ *n.* 国籍
link /lɪŋk/ *v.* 联合，联合
well-structured /welˈstrʌktʃəd/ *adj.* 结构良好的

collect /kəˈlekt/ *v.* 收集，搜集
state /steɪt/ *v.* 陈述，声明
character /ˈkærəktə(r)/ *n.* 文字
date /deɪt/ *n.* 日期
take up 占用，占据

True or false.
1. In the past, people retrieved information by hand.
2. The card filing system needs a lot of room to store data.
3. Computer databases cannot store as many data as card filing systems.
4. The data in databases can only be displayed in the form of characters and numbers.
5. When you buy database software, all the data you need are already set up in the system.

Notes

Data 数据：数据是指存储在某一媒体上并能识别的物理符号，数据的概念包

107

括两个方面，其一是描述事物特性的数据内容，其二是存储在某一种媒体上的数据形式。数据不仅包括由数字、字母、文字和其他特殊字符组成的文本形式，而且还包括图形、图像、动画、影像、声音等多媒体形式，但使用最多的仍然是文字数据。

Access：是一个功能强大的Windows应用程序，它使数据库管理系统的功能兼备了Microsoft Windows的通用性。Access集成在Office中，在这种软件环境下，Access可以充分发挥其功能。

Do You Know ?

数据库的发展

第一阶段（20世纪70年代），其RDBMS（Relational Database Management System）仅支持关系数据结构和基本的关系操作（选择、投影、连接），如：DBASE。

第二阶段（20世纪80年代），其产品对关系操作的支持已经比较完善，但是对数据完整性的支持仍然较差。此时，SQL语言已经成为关系数据库的标准。

第三阶段（20世纪90年代），其产品加强了数据完整性和安全性。完整性的控制在核心层实现，克服了在工具层的完整性上可能存在"旁路"的弊病。

目前，SQL Server、Oracle等数据库系统都支持业务云化，注重大数据的智能分析服务。

Unit 7 Database

Language Practice

I. Fill in the blanks with the missing letters of the words according to the pictures.

1. a h_t_l

3. a gr_cery st_re

2. a s_perm_ _ket

4. a r_c_ption_st

6. a m_le

5. a cl_n_c

8. a p_t_ _nt

10. d_t_

7. d_tab_se s_ftw_re

9. a card rec_ _d

109

II. Complete the sentences with the words practiced above.

1. Nowadays, _____ are almost replaced by database files.
2. Do you still remember the _____ when we first met?
3. I picked up some eggs in the _____ around the corner.
4. If you have a fever, you may go to the _____ to do a check-up.
5. You may ask the _____ there when the next plane to New York will take off.
6. It is easy for you to make a report with the _____.
7. The Johnson's _____ infant is so lovely.
8. The _____ is now open to guests.
9. There are always so many _____ in the Central Hospital.
10. Another _____ has been set up in our neighborhood.

III. Fill in the blanks with the words in the box.

form	record	query	table	field

Microsoft Access is an example of a database program. Database programs can be used to store and manage information.

1. database — a place where lots of information is stored in a logical way.
2. _____ — all the information stored about one person, place or thing.
3. _____ — one piece of a record, for example the person's family name or a car's registration number.
4. _____ — the most common way to view records in a database. Columns are fields and rows are records.
5. _____ — the cell used to enter data into a record.
6. report – the most common way to view information in a database.
7. primary key — a field used to identify each individual record.
8. _____ — a way of searching for information in a database.

Unit 7 Database

IV. Match the ways of creating a new database with conditions.

When you open Access, the new tab provides several ways that you can create:

1. A blank database _____
2. A template that is installed with Access _____
3. A template from Office.com _____

Conditions:

a. If you are starting a new project and it is a head start.
b. If you have very specific design requirements or have existing data.
c. If you want to find many more templates on Office.com, besides the templates that come with Access.

V. Change the order of the sentences. You can also do this with the help of your computer.

How to create a blank database

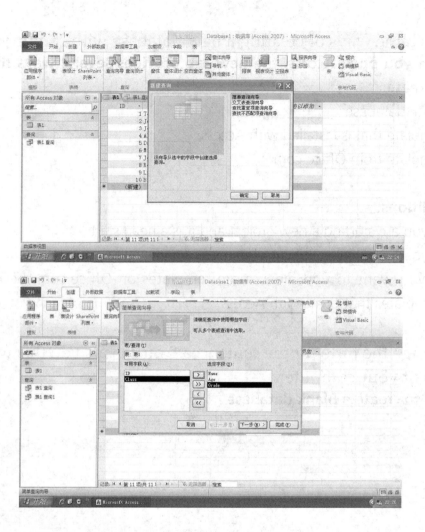

1. Click "Create".
2. Begin typing to add data, or you can paste data from another source.
3. On the File tab, click "New", and then click "Blank Database".
4. In the right pane, under "Blank Database", type a file name in the File Name box.

VI. Sentence completion.

1. Every time a customer _____ the hotel, the information is entered into the database. (每次顾客入住酒店，其信息都会被录入数据库。)
2. We have _____ a database of all our patients. (我们已经建立了所有

患者的数据库。)

3. The old sofa _____ a large amount of space in the room. (旧沙发占据了房间内很大的空间。)
4. The data in a database can be in the form of numbers, characters, dates _____. (数据库中的数据可以是数字、文字、日期等形式。)
5. Each row in a relational database holds the same _____ information. (关系型数据库的每行中存有同类信息。)

Writing

Fill in the blanks with the information you have got from the texts.

A database is an application that manages _____ and allows fast _____ and retrieval of data. There are different types of databases but the most popular is a _____ database. _____ is a simple programming language that is used in relational databases. Both _____ relational databases and _____ relational databases exist.

Grammar

	Passive Voice (被动语态)
be + (not) + done	Notebook computers **are** also **called** laptop computers. All the data **are linked** in a well-structured way, and can be retrieved easily and quickly. The girl **is not named** Lily but Sarah. **Is** New York also **called** Big Apple?
can / may / must / should (not) be + done	All the data are linked in a well-structured way, and **can be retrieved** easily and quickly. Books in the reading-room **should not be taken** out of the library. This book **must be returned** tomorrow, for it is due.

I. Multiple choices.

1. In some schools, computer skills _____.
   A. teaches   B. are taught   C. taught
2. The computing system will be further attacked unless some measures _____.
   A. will be taken   B. are taken   C. were taken
3. Microsoft _____ in 1975.
   A. was founded   B. founded   C. found
4. Lots of information _____ in a logical way when a database program is used.
   A. store   B. would store   C. would be stored
5. This file _____ before it is closed.
   A. ought to save   B. should be saved   C. should save

II. Fill in the blanks with the correct forms of the verbs in the brackets.

1. He asked his mother for some money to buy a new laptop, but _____ (give) a good scolding.
2. I'm afraid you must _____ (file) the document right now.
3. Mr Black _____ (use) Word to store data but his workmate _____ (not use) the software very often.
4. The computer in our office _____ (break) down. That is why we _____ (use) the manual system at present.
5. Databases are widely _____ (install) in office computers.

Word formation — Derivation (派生法)

后缀-ist加在名词后，表示与某种活动或事物有关的人；后缀-al加在名词后构成形容词。

Unit 7 Database

Noun	Suffix	Noun / Adjective
emotion	-al	emotional
relation	-al	relational
environment	-al	environmental
science	-ist	scientist
art	-ist	artist

III. Complete the sentences with the derivative words above.
1. Sally is very _____; she cried even when she cannot solve the problem on her computer.
2. _____ are always respected by people.
3. Nowadays, _____ also use computers to create their works.
4. The most popular database is _____ database.
5. Nobody knows how to solve this _____ problem.

Game

Across

1. Jill _____ed/d a program for the company, which brought fame to her.
2. You should _____ data before you set up a data file.
3. The laptops sold here are _____ed from America.
4. A database is an application that allows fast _____ of data.
5. Johnson's school offers _____ training to the pupils.

115

Down

1. Have you installed a database _____ in your computer?
2. All the data are _____ed in a well-structured way in computers.
3. Oracle is a popular _____ relational database.
4. A grocery store owner can use a database system to keep an _____ of his store.

 Fun Time

Computer Humour

Customer: How much do Windows cost?

Tech Support: Windows costs about $100.

Customer: Oh, that's kind of expensive. Can I buy just one window?

 Project

Communicate with others!
Step 1: Make a list of occupations and industries, which find databases useful.
Step 2: Within each group, talk about what they use databases for.
Step 3: Report one of the applications of databases in detail in class.

 Self-checklist

根据实际情况，从A、B、C、D中选择合适的答案：A代表你能很好地完成该任务；B代表你基本上可以完成该任务；C代表你完成该任务有困难；D代表你不能完成该任务。

A B C D

☐ ☐ ☐ ☐ 1. 能掌握并能运用本单元所学重点句型、词汇和短语。
☐ ☐ ☐ ☐ 2. 能理解并正确模仿听说部分的句子，正确掌握发音及语调。
☐ ☐ ☐ ☐ 3. 能模仿句型进行简单的对话。
☐ ☐ ☐ ☐ 4. 能读懂本课的短文，并正确回答相关问题。
☐ ☐ ☐ ☐ 5. 能掌握被动语态的用法。
☐ ☐ ☐ ☐ 6. 能掌握以-ist, -al为后缀的派生词的构词方法。
☐ ☐ ☐ ☐ 7. 能用课文中学习的词组造句。
☐ ☐ ☐ ☐ 8. 能向同学介绍数据库及其种类等基本知识。

Unit 8 PowerPoint

Unit Goals

In this unit, you will be able to
- understand listening materials about PowerPoint;
- talk about PowerPoint;
- understand articles about PowerPoint presentation;
- write a short paragraph about PowerPoint;
- demonstrate your knowledge about PowerPoint.

Lead-in

Picture matching.

1. making a presentation   2. slide   3. projector
4. whiteboard   5. overhead projector   6. pointer

  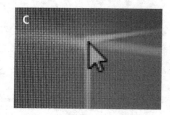

Unit 8 PowerPoint

Listening and Speaking

Listen and complete.

a. pictures and videos
b. make a presentation
c. sales profits
d. introduce our products

What Is PowerPoint?

Tom, a teenager, is busy working with his computer.

❶ Mom: Tom, what are you doing there with PowerPoint?
    Tom: I'm doing my homework. Our history teacher asked us to _____ for our group's work.
    Mom: So what are you going to do with pictures?
    Tom: Well, I am collecting some pictures about the Silk Road in the Tang Dynasty.

Andy is a college student. He is doing a survey for his assignment. He is now talking with Jack, a salesman.

❷ Andy: Hello, sir. What do you do?
    Jack: I am a salesman. I sell computer systems to business people.
    Andy: Do you use PowerPoint quite often?
    Jack: Yes. I usually use PowerPoint to _____. The

119

presentation helps people know how the systems can improve their work.

Andy is now talking with Ms Wang, a university teacher.

3   Andy: Excuse me, madam. What do you do?
Ms Wang: I'm a teacher. I teach English in a college.
Andy: Do you use PowerPoint in your teaching?
Ms Wang: Definitely. PowerPoint can integrate sound, images, _____. It's easy to use and it can bring life to my class.

Andy is now talking with Mr William, a sales manager of a company.

4   Andy: Good morning, sir. What are you doing with PowerPoint?
Mr Williams: I'm preparing a presentation for the annual meeting of the company.
Andy: What are you going to do with it?
Mr Williams: I'm going to show the Board of Directors _____ within the last 10 months.

Role Play

A: Hello, sir / madam. ... What are you doing with PowerPoint?
B: Well, I'm preparing a presentation / preparing a lesson ...

Unit 8 PowerPoint

slide /slaɪd/ n. 幻灯片
whiteboard /'waɪtbɔːd/ n. 白板
overhead /ˌəʊvə'hed/ adj. 头顶的
dynasty /'dɪnəsti/ n. 朝代
integrate /'ɪntɪɡreɪt/ v. 使结合成为整体
image /'ɪmɪdʒ/ n. 图像

annual /'ænjʊəl/ adj. 每年的

be busy doing sth 忙于做某事
Silk Road 丝绸之路
do a survey 做调查
be easy to use 易于使用

Reading

Text A

Pre-reading activities.

1. When you are asked to make a presentation, which of the following would you need? Tick your choices.
   - ☐ PowerPoint slides
   - ☐ hand-drawn slides
   - ☐ mechanically typeset slides
   - ☐ a blackboard
   - ☐ a whiteboard
   - ☐ an overhead projector
   - ☐ a video projector
   - ☐ a computer
   - ☐ a pointer

2. Have you ever attended a PowerPoint presentation? If yes, what do you think of it?

> **Reading Strategy**
>
> 记笔记 (Note-taking) 记笔记是将所学知识进行适当的记录以促进理解和记忆的方法。当学习比较复杂的内容，需要把握内容的要点时，使用记笔记的策略能产生最大的积极作用。另外，记笔记时如果能够同时在心里进行归纳，而不是简单地记下所看到或听到的内容，那么效果会更佳。

PowerPoint

The past decades has seen many new developments in computer technology. These advancements have really made a difference in our lives. For example, if you have to make a presentation at work for certain projects or at school for some classes, you can do it with the help of PowerPoint, a presentation program.

PowerPoint presentations consist of a number of individual pages or "slides". Slides may contain text, graphics, movies, and other objects, which may be arranged freely on the slide. PowerPoint not only increases the overall outlook, but also brings the audience in.

The presentation can be printed, displayed live on a computer, or navigated through at the command of the presenter. For larger audiences the computer display is often projected by using a video projector.

The use of presentation software can save a lot of time for people who otherwise would have used other types of visual aid — hand-drawn or mechanically typeset slides, blackboards or whiteboards, or an overhead projector.

Unit 8 PowerPoint

Words & Expressions

decade /'dekeɪd/ n. 十年
technology /tek'nɒlədʒi/ n. 技术
advancement /əd'vɑːnsmənt/ n. 进步
consist /kən'sɪst/ v. 组成
individual /ˌɪndɪ'vɪdʒuəl/ adj. 单独的；个人的
contain /kən'teɪn/ v. 包含
graphics /'græfɪks/ n. 图案
arrange /ə'reɪndʒ/ v. 安排
overall /ˌəʊvər'ɔːl/ adj. 总体的，全面的
outlook /'aʊtlʊk/ n. 外观

audience /'ɔːdiəns/ n. 观众，听众
navigate /'nævɪgeɪt/ v. 导航
visual /'vɪʒuəl/ adj. 视觉的
mechanically /mə'kænɪkli/ adv. 机械地

make a difference 使不同
with the help of 在……的帮助下
consist of 由……组成
not only ... but also ... 不仅……而且……
bring ... in 吸引；带入

Fill in the blanks with the words from the text to summarise Text A.

___1___ has improved a lot in the past decades. For example, ___2___ is one of its advancements. Its presentations have many ___3___ which contain ___4___, graphics, movies, and other objects. When making a presentation, one can project the computer ___5___ with a video projector. Using presentation ___6___ instead of other visual aids can save people's time.

Text B

Pre-reading questions.
1. Have you ever used PowerPoint? If yes, for what?
2. What are the steps to make a presentation with PowerPoint?

How to Create PowerPoint

Microsoft PowerPoint gives you more ways to create and share dynamic presentations with your audience than ever before. No matter what the topic is, a PowerPoint presentation can satisfy your needs when you communicate with your audience. Here are some steps to get you started.

1. Start with a template (background) on which your text, pictures, graphs or videos will be shown. (FILE→NEW→FROM TEMPLATE)
2. Choose the format of your slide, for instance, if you want to have a single slide with text on the right and a picture on the left. Choose the text or content layout.
3. If you want to type text, choose a "slide layout" with a text box. Same process for a picture, movie or sound file.
4. To insert anything (text, charts, etc), simply click on the designated box to activate it and start inserting.
5. To play the presentation, go to: VIEW→SLIDE SHOW or press "F5".

Words & Expressions

dynamic /daɪˈnæmɪk/ adj. 动态的
topic /ˈtɒpɪk/ n. 主题
communicate /kəˈmjuːnɪkeɪt/ v. 交流
template /ˈtempleɪt/ n. 模板
instance /ˈɪnstəns/ n. 实例
format /ˈfɔːmæt/ n. 格式
layout /ˈleɪaʊt/ n. 布局，安排

sample /ˈsɑːmpl/ n. 样本
chart /tʃɑːt/ n. 图表
designated /ˈdezɪɡneɪtɪd/ adj. 指定的
activate /ˈæktɪveɪt/ v. 激活

satisfy one's needs 满足某人的需求

Unit 8 PowerPoint

I. Put the following sentences in correct order according to the steps listed in the text.
1. If you want to insert text, charts, etc, click on the designated box to activate it and start inserting.
2. Choose a "slide layout" with a text box to type text. Same process for a picture, movie or sound file.
3. Choose a template where the text, pictures, graphs or videos can be shown.
4. Go to: VIEW → SLIDE SHOW or press "F5" to play the presentation.
5. Choose a format for each of your slide.

II. Make a presentation with PowerPoint following the steps introduced in Text B.

Notes

PowerPoint：和Word、Excel等应用软件一样，都是Microsoft公司推出的Office系列产品之一，主要用于演示文稿的创建，即幻灯片的制作，它可有效帮助演讲、教学、产品演示等。

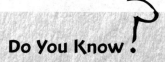

大屏幕投影机

　　大屏幕投影机可将人们所需交流的资料、图片、实物、动态画面等交流信息和VCD、LD或DVD、VCR等视频信号和计算机信号播放成大屏幕动态画面和计算机画面，其尺寸远远大于电视机、电脑显示器，故称大屏幕投影机。它适用于学术会议、大专院校和中小学的多媒体教学、电子化培训、家庭娱乐、商务演示及工业监控等。大屏幕投影机可分为胶片式投影机、视频投影机及多媒体投影机。

Language Practice

I. Fill in the blanks with the missing letters of the words according to the pictures.

2. a wh_teb_ _ _d

1. te_pl_te

3. a ch_ _t

5. te_t

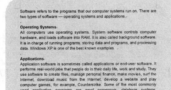

4. pr_sent_ tion

6. gr_ph_cs

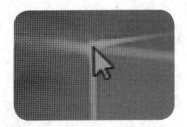

7. a proj_ct_ _

8. a p_inter

II. Complete the sentences with the words practiced above.
1. The students could not see the _____ you typed on the screen clearly.

2. In some cases, the rights to these photos or _____ are wholly or partly owned by others.
3. We can illustrate our data with the column _____.
4. To make a _____, David needs a whiteboard.
5. A figure was shown through the overhead _____.
6. A _____ is a convenient handy tool for presentations, teaching, meetings and speeches.
7. Nowadays _____ are replaced by smart boards little by little.
8. You may find a lot of beautiful _____ online to make your presentation.

Ⅲ. Give a proper title to each part according to the steps.

> Save a presentation
> Find and apply a template
> View a slide show
> Create a presentation
> Insert a new slide
> Open a presentation

1. ( )
PowerPoint allows you to apply built-in templates. To find a template in PowerPoint,
1) on the File tab, click "New".
2) under "Available Templates" and "Themes", click "Sample Templates", click the template that you want, and then click "Create".

2. ( )
Click the File tab, and then click "New". Apply a template or theme from those built-in ones with PowerPoint.

3. (   )
1) Click the File tab, and then click "Open".
2) In the left pane (方框) of the Open dialogue box, click the drive or folder that contains the presentation that you want.
3) In the right pane of the Open dialogue box, open the folder that contains the presentation.
4) Click the presentation, and then click "Open".

4. (   )
1) Click the File tab, and then click "Save As".
2) In the File name box, type a name for your PowerPoint presentation, and then click "Save".

5. (   )
To insert a new slide into your presentation, on the Home tab, in the Slides group, click the arrow below "New Slide", and then click the slide layout that you want.

6. (   )
To view your presentation in Slide Show from the first slide: on the Slide Show tab, in the Start Slide Show group, click "From Beginning".
To view your presentation in Slide Show from the current slide: on the Slide Show tab, in the Start Slide Show group, click "From Current Slide".

IV. Sentence completion.

1. This kind of template can _____. (这种模板可以满足大多数人的要求。)
2. You can _____ text, graphics, etc. _____ a slide. (你可以在幻灯片上插入文本、图表等。)
3. Computers _____ in people's lives. (计算机使人们的生活有

了很大不同。)

4. Students are learning to make PowerPoint _____ of the teacher. (同学们正在老师的帮助下学习制作演示文稿。)
5. PowerPoint can _____ be printed, _____ displayed live on a computer. (演示文稿不仅可以打印，还可以在计算机上现场演示。)

Writing

Write a short paragraph (at least 5 sentences) about PowerPoint by answering the following questions.

What is PowerPoint?
What is it used for?
What does it contain?
What good can it do to us?

Grammar

Adverbial Clause (状语从句) (1)		
of time	when, while, as, before, after, since, until, till, as soon as ...	A PowerPoint presentation can satisfy your needs **when** you communicate with your audience.
of condition	If, unless, as long as ...	Choose the format of your slide, for instance, **If** you want to have a single slide with text on the right and a picture on the left. **If** you want to type text, choose a "slide layout" with a text box as shown in the sample template picture.

129

(续表)

of purpose	so as to, so that, in order that, lest ...	Click on the designated box **so that** you may activate it and start inserting. Go to: VIEW→SLIDE SHOW or press "F5" **so that** you can play the presentation.
of comparison	than, as ... as, the ... ,the ...	Microsoft PowerPoint gives you more ways to create and share dynamic presentations with your audience **than** ever before.
of concession	No matter (what / how / which / ...), though, although, (how, what, who, ...)+ever, ...	**No matter what** the topic is, a PowerPoint presentation can satisfy your needs when you communicate with your audience.

I. Multiple choices.

1. The students listened carefully _____ Bill made the PowerPoint presentation.

    A. when　　　　　　B. as soon as　　　C. before

2. Peter chose to make the presentation in class _____ he was nervous.

    A. no matter what　　B. however　　　　C. although

3. _____ Tony has to make a presentation at school, he would do it with the help of PowerPoint.

    A. Unless　　　　　B. If　　　　　　　C. After

4. Lily chose a "slide layout" with a text box _____ she can type text.

    A. so that　　　　　B. in that　　　C. in order to

5. As a programmer, Linda knows the PowerPoint software _____ Lucy.

    A. better than　　　　B. as better as　　　C. good than

130

II. Fill in the blanks with the correct words in the box.

| when | If | as ... as | to | although |

_____ you work with others on presentations and projects, PowerPoint is the perfect tool for you. _____ you are working with a team on an important presentation, PowerPoint gives you the power to work more easily. PowerPoint delivers new and improved tools _____ add power to your presentations. _____ PowerPoint allows you to apply built-in templates, you can apply your own custom templates. Exciting new audio and visual capabilities help you tell a cinematic (精彩的，电影般的) story that is _____ easy to create _____ it is powerful to watch.

Word formation — Derivation (派生法)

某些动词后面加上-ment, -ence, -ion (-tion, -ation) 等后缀可构成名词，如：argument, realisation。

III. Add –ence, -ment or -ion (-ation, -tion) to the following verbs to form nouns.

verb	noun	verb	noun
develop		advance	
navigate		integrate	
differ		assign	
present		arrange	
collect		introduce	

Game

Across

1. From the _____ of the template, we can learn that Tony has artistic taste.
2. A _____ is used to make new pages with a similar design, pattern, or style.
3. Lily is preparing a presentation for the _____ meeting of the company.
4. The student is asked to deliver a twenty-minute _____ in class.
5. Don't be nervous though you have a large _____.

Down

1. Sue's company is producing books in all kinds of different _____s.
2. The film is a _____ art.
3. There is little space on the slide for the _____.
4. When you _____ with your audience, a PowerPoint presentation is a great help.
5. Nowadays _____s are widely used in classroom teaching.

Fun Time

Heaven or Hell?

A dying programmer found himself in front of a committee that decides whether a person should go to Heaven or Hell. He was told that he could have a look of Heaven and Hell before he made the decision.

He was then taken by an angel to a place with a sunny beach, volleyball and music, where everyone was having a great time.

"Wow!" he exclaimed. "Heaven is great!"

"Wrong," said the angel. "That was Hell. Want to see Heaven?"

"Sure!" So the angel took him to another place, where some people were feeding dead pigeons.

"This is Heaven?" asked the programmer.

"Yup," said the angel.

"Then I'll take Hell." Immediately he found himself in red-hot lava, with some miserable people around him. "Where's the beach? The music? The volleyball?" he screamed to the angel.

"That was the demo," she replied as she went away.

Project

Communicate with others!

Step 1: Within each group, talk about the benefits of using PowerPoint in presentation.

Step 2: Within each group, decide on a topic and make some PowerPoint slides for it.

Step 3: Ask one member from each group to make a PowerPoint presentation on the topic.

Self-checklist

根据实际情况，从A、B、C、D中选择合适的答案：A代表你能很好地完成该任务；B代表你基本上可以完成该任务；C代表你完成该任务有困难；D代表你不能完成该任务。

A B C D

☐ ☐ ☐ ☐ 1. 能掌握并能运用本单元所学重点句型、词汇和短语。

☐ ☐ ☐ ☐ 2. 能理解并正确模仿听说部分的句子，正确掌握发音及语调。

☐ ☐ ☐ ☐ 3. 能模仿句型进行简单的对话。

☐ ☐ ☐ ☐ 4. 能读懂本课的短文，并正确回答相关问题。

☐ ☐ ☐ ☐ 5. 能掌握时间、条件、目的、比较和让步状语从句的用法。

☐ ☐ ☐ ☐ 6. 能掌握时间、条件、目的、比较和让步状语从句引导词的用法。

☐ ☐ ☐ ☐ 7. 能掌握以-ment, -ence, -ion (-tion, -ation) 为后缀的派生词的构词方法。

☐ ☐ ☐ ☐ 8. 能用课文中学习的词、词组造句。

☐ ☐ ☐ ☐ 9. 能向同学介绍演示文稿软件及制作等相关知识。

Unit 9

Desktop Publishing Software

Unit Goals

In this unit, you will be able to
- understand listening materials about desktop publishing software;
- talk about desktop publishing software;
- understand articles about desktop publishing software;
- write a short paragraph about desktop publishing software;
- demonstrate your desktop publishing software knowledge.

Lead-in

Picture matching.

1. QuarkXPress
2. Desktop Publishing Software
3. brochure
4. InDesign

Listening and Speaking

Listen and complete.

a. I have included lots of pictures and interesting stories of some famous basketball players.
b. I've just opened a fast food restaurant.
c. Sure, it's a piece of cake.
d. What a lovely dog!

How Can the Software Help You?

In a computer lab of a training school, some students are using Desktop Publishing Software to create their own works of art. A technician is walking around to see if they need any help.

He is now talking with Paul, a university student.

❶ Technician: Hi, young man. Can I help you?
  Paul: Yeah. I've just lost my dog! I'm going to make some posters with his pictures on.
  Technician: _____ I hope you can have him back soon.

He is talking to Nick, a basketball fan.

❷ Technician: Hello, Nick. Your screen looks colourful and beautiful. What are you designing?
  Nick: I'm creating a basketball webpage with Publisher.
  _____
  Technician: I am a basketball fan, too. And I have collected a great many pictures of Yao Ming. Do you want to have more information about him?
  Nick: Thanks. You're really a great help.

He is now talking with a man in his late thirties.

❸ Technician: Excuse me. What can I do for you?

Unit 9 Desktop Publishing Software

        Man: I'm making the new menu for my restaurant using Publisher. I want it to look eye-catching.
Technician: Do you own a restaurant?
        Man: Yup. _____ It's just across the street. Why not go and try our food?
Technician: Sure, I will. Good luck!

A pretty girl comes into his sight. Her name is Susan.

4   Susan: Hi, could you do me a favour?
Technician: Ready to serve you.
    Susan: My boss asked me to create a set of business cards, envelopes and letter heads for our newly set-up firm. But I don't know how to use Publisher. Could you give me a hand?
Technician: _____

Later all the work is finished.
    Susan: I owe you a favour.
Technician: It is my pleasure, Miss.

Role Play

A: Hi, can I help you? / what can I do for you?
B: I'm creating / making some posters / a basketball web page / the new menu for my restaurant / a set of business cards ... with Publisher / using Publisher ...

A: ...
B: Thanks a lot / Thank you so much. / You're a great help. / I own you a favour.
A: You're welcome. / It's my pleasure.

Words & Expressions

brochure /ˈbrəʊʃə(r)/ *n.* 小册子
include /ɪnˈkluːd/ *v.* 包括
technician /tekˈnɪʃn/ *n.* 技术人员
poster /ˈpəʊstə(r)/ *n.* 海报
web page /ˈweb peɪdʒ/ *n.* 网页
eye-catching /ˈaɪ kætʃɪŋ/ *adj.* 引人注目的
sight /saɪt/ *n.* 看见

firm /fɜːm/ *n.* 公司
pleasure /ˈpleʒə(r)/ *n.* 乐事

Good luck! 祝你好运！
do sb a favour 帮某人忙
a piece of cake 很容易，小菜一碟
owe sb a favour 欠某人一个情

Reading

Text A

Pre-reading activities.

1. Which would you use when you are creating documents, word processor or desktop publishing software? State your reasons.

2. What desktop publishing software do you know? Tell your classmates about it.

3. When you buy desktop publishing software, what are the things you consider?

Reading Strategy

找主题句 Reading for Topic Sentences 在阅读英语文章时，找到并仔细研读主题句对理解篇章有很大作用。大多数情况下，主题句都出现在比较重要的位置，如文章的开头或结尾。如果一篇文章包括多个段落，一般来说每个自然段的首句是主题句。文章主题句之外的内容一般都是对主题句进一步加以解释、补充说明或列举事实等。

Choosing Desktop Publishing Software

If the documents you are creating are becoming too complex, you may want to use desktop publishing (DTP) software so that you can work easily. Software like Adobe InDesign and QuarkXPress are specifically designed to make it easy to create documents that combine text and graphics in complex layouts. Many people have avoided using DTP software because they were afraid it was too difficult to learn and use. Although many graphic artists and desktop publishers do use these programs, you do not need a design degree to understand the basics of laying out a page. You may find that for many tasks, these programs are actually easier to use than your word processor.

When you are shopping for desktop publishing software, think about the documents you want to create. All desktop publishing software includes powerful typographic and graphic features, but, as is known, certain products excel at creating certain types of documents. For example, Adobe InDesign is good at creating short documents like brochures, while Adobe FrameMaker is designed for long documents like technical manuals.

Words & Expressions

complex /ˈkɒmpleks/ adj. 复杂的
specifically /spəˈsɪfɪkli/ adv. 特意；专门地
combine /kəmˈbaɪn/ v. 使结合
avoid /əˈvɔɪd/ v. 避免
lay /leɪ/ v. 放置，平铺
typographic /ˌtaɪpəˈɡræfɪk/ adj. 印刷上的，排字上的
excel /ɪkˈsel/ v. 优于，擅长

technical /ˈteknɪkl/ adj. 技术的

avoid doing 避免做某事
be afraid (that) 害怕，担心
too ... to 太……以至于不……
as is known 众所周知
excel at (在某方面) 出色

Which are topic sentences and which are supporting sentences? Write T for topic sentences and S for supporting sentences in the boxes.

1. ☐ If the documents you are creating are becoming too complex, you may want to use desktop publishing (DTP) software so that you can work easily.
2. ☐ Software like Adobe InDesign and QuarkXPress are specifically designed to make it easy to create documents that combine text and graphics in complex layouts.
3. ☐ Although many graphic artists and desktop publishers do use these programs, you do not need a design degree to understand the basics of laying out a page.
4. ☐ When you are shopping for desktop publishing software, think about the documents you want to create.
5. ☐ For example, Adobe InDesign is good at creating short documents like brochures, while Adobe FrameMaker is designed for long documents like technical manuals.

Text B

Pre-reading questions.
1. Are there any differences between word processor and desktop publishing software? Discuss with your partner.
2. Do you know what the letters DTP stand for?

How DTP Works

In a desktop publishing (DTP) program, you place elements on the page and move them around freely. Repeating elements, such as a logo, can be placed on master pages. You can set up as many different master pages as you wish and apply one to any page in your document to quickly add or change formatting.

You can create a new publication with a template. Wherever you plan to reuse a layout for a document, you can save the document as a template.

In a desktop publishing program, you add text by importing it from your word processor or by typing it in directly. Text is generally placed into a box called a frame. You can create the layout first, link the frames together, and flow the text into them later.

Once you have your text in your document, you format it by applying attributes such as font settings or colours. You can apply the formatting to each individual piece of text or using paragraph styles. With styles, your text formatting remains consistent throughout the document.

Words & Expressions

element /ˈelɪmənt/ n. 成分；要素
logo /ˈləʊɡəʊ/ n. 标志，商标
master /ˈmɑːstə(r)/ n. 母版
publication /ˌpʌblɪˈkeɪʃn/ n. 出版物
generally /ˈdʒenrəli/ adv. 通常，大体上
frame /freɪm/ n. 框架，文本框
attribute /ˈætrɪbjuːt/ n. 属性，特性

format /ˈfɔːmæt/ v. 排版，版式设计……
n. 版面
consistent /kənˈsɪstənt/ adj. 一致的
throughout /θruːˈaʊt/ prep. 自始至终
apply to 应用于

Multiple choices.

Choose the correct answers for the following questions according to Text B.

1. What can be placed on a master page?
   A. Elements.　　　　B. Logos.　　　　C. Programs.
2. How many times can you use a template?
   A. Many times.　　　B. Two times.　　　C. One time.
3. How do you add text in a desktop publishing program?
   A. By importing it from the word processor.
   B. By typing it in directly.
   C. Both A and B.
4. Text is usually placed in a _____.
   A. frame　　　　　　B. layout　　　　　C. processor
5. What make the text formatting consistent throughout the document?
   A. Settings.　　　　B. Colours.　　　　C. Styles.
6. This text is about _____.
   A. how to develop a desktop publishing program
   B. how to use a desktop publishing program
   C. how to create a new publication with a template

Notes

Desktop Publishing Software 桌面排版软件：桌面排版软件具有排版、设计、色彩和图形处理、专业制图、文字处理及印前作业功能。它可用于宣传手册、杂志、书本、广告、商品目录、报纸、包装、技术手册、年度报告、贺卡、传单、建议书等的设计、排版和制作。

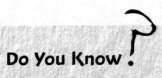

常用的排版软件

目前常用的排版软件有Adobe公司的InDesign，Quark公司的QuarkXpress，北大方正公司的FIT（飞腾）和Microsoft公司的Publisher。Publisher是Microsoft Office组件之一，是一款入门级的桌面出版应用软件。InDesign的特点有：能输出PDF及HTML格式文件，图层管理、色彩管理功能强，图文链接、表格制作功能独特。QuarkXpress的特点有：可自动备份并存储，具备组页功能，可输出EPS格式文件，支持渐变填充图形等。FIT的功能特点有：中文处理功能较强，能满足中文的各种编排要求；图形绘制功能强、预置底纹图案多、变换功能强。

Language Practice

I. Fill in the blanks with the missing letters of the words according to the pictures.

1. man_al

2. a l_go

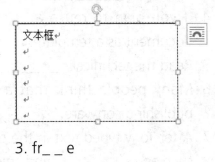
3. fr_ _ e

4. f_nt s_tting

5. a mas_ _ r page

6. lay_ _ t

143

7. a doc_ me_ t   8. word pro_ _ ssor

II. Complete the sentences with the words practiced above.

1. The secretary typed the text into the _____.
2. A _____ can be applied to any page in your document to quickly add or change formatting.
3. There is a _____ of the college on each page of the template.
4. If you plan to reuse a _____ for a document, you can save the document as a template.
5. Read the technical _____ before starting the printer.
6. Many people think that a _____ is easier to use than desktop publishing software.
7. After Tony typed text in the document, he formatted it by applying attributes such as _____ and colours.
8. Desktop publishing software can be used to create complex _____.

III. Complete the passage with the words in the box.

documents	*desktop publishing*	*tools*	*PagePlus*	*pages*

PagePlus from Serif is _____ software for users from all levels of computer experience. Once you become familiar with _____, it is quite easy to use.

The top toolbar includes choices to adjust text and page qualities as well as open, save, and print _____. The left toolbar provides _____ for working with text, images, shapes, drawing and effects. The right toolbar is for changing the attributes of documents, viewing _____, colour schemes (配色方案) and adding object styles.

IV. Sentence completion.

1. The logo can be _____ the top of the template. (标志可以放在模板的上方。)
2. There is a lot of desktop publishing software, _____ QuarkXpress and InDesign. (有很多种桌面排版软件，比如QuarkXpress、InDesign等。)
3. You can set up more than one _____. (你可以建立不止一个母板。)
4. Many people _____ that desktop publishing software is difficult to learn and use. (许多人担心桌面排版软件难学、难用。)
5. Adobe InDesign _____ creating short documents like brochures. (Adobe InDesign 在创建短文件，比如小册子方面很出色。)

Writing

Write a short paragraph (at least 5 sentences) about DTP by answering the following questions.
1. When do people use desktop publishing software?
2. What does desktop publishing software include?
3. Where do people place elements in a desktop publishing program?
4. Where and how do people add text in a desktop publishing program?
5. How do people format text in the document?

Grammar

Adverbial Clause (状语从句) (2)		
of manner	as, as if, as though	**As** is known, certain desktop publishing products excel at creating certain types of documents.
of result	so that, so ... that ... , such ... that	Johnson used desktop publishing (DTP) software **so** skillfully **that** he finished his work earlier than the other.
of place	where, wherever	**Wherever** you plan to reuse a layout for a document, you can save the document as a template.
of cause	because, as, since, for, ...	Many people do not use DTP software **because** they think it is too difficult to learn and use.

I. Multiple choices.

1. Desktop publishing software is _____ easy to learn _____ Bill could use it skillfully now.
   A. such ... that   B. so ... that   C. as ... as
2. The student applied the master page _____ the teacher told him.
   A. as   B. as if   C. as though
3. Please set up master pages _____ you can quickly add or change formatting.
   A. because   B. as   C. so that
4. I will start making the brochure _____ I stopped yesterday.
   A. where   B. so that   C. since

5. Lucy likes the desktop publishing software _____ it is friendly for users from all levels of computer experience.
   A. as if   B. where   C. because

II. Fill in the blanks with the correct words in the box.

| as | because | so ... that | where |

We really like Print Artist Platinum _____ it includes professional-looking templates, as well as great tools for working with larger documents such as books. Apply it _____ you want, you will be able to create every type of project. Find the template you need, throw a few graphics in there _____ you are instructed and you have a professional project that is ready to publish. The results you get will be _____ impressive _____ you would be glad to use it.

Word formation — Derivation (派生法)

在派生构词法中，re-作为前缀，放在动词、名词、形容词前，表示"又"、"再"、"重新"、"回复"，如：reuse, reaction, rebuilt 等。

III. Add prefix re- to the following words to form derivative words and tell their Chinese meanings.

verb		noun		adjective	
use		adjustment		built	
activate		development			
arrange		education			
apply					

Game

Across

1. What you're saying now is not _____ with what you said last week.
2. All the _____s are put on the page in a DTP program.
3. Chris _____ed answering my questions, for she didn't review what we had learnt well.
4. Certain desktop publishing software _____s at creating certain types of documents.

Down

1. Celina created a _____ online for her class.
2. InDesign makes it easy to create documents that _____ text and graphics in complex layouts.
3. Billy designed a new _____ for the company.
4. Elaine made a _____ on her computer and she is going to print it out today.
5. Once a _____ page is made, it can be applied to all the slides.
6. _____ is a complete set of letters and numbers in one size and style used for computer documents.

Unit 9 Desktop Publishing Software

Fun Time

Software Engineer

A man was crossing a road when a frog jumped out to him and said: "If you kiss me, I'll turn into a beautiful princess." He bent over, picked up the frog and put it in his pocket.

The frog spoke up again and said: "If you kiss me and turn me back into a beautiful princess, I will stay with you for one week."

The man took the frog out of his pocket, smiled at it and returned it to the pocket.

The frog then cried out: "If you kiss me and turn me back into a princess, I'll stay with you and do anything you want." Again the man took the frog out, smiled at it and put it back into his pocket.

Finally, the frog asked: "What's the matter? I've told you I'm a beautiful princess, and I'll stay with you for a week and do anything you want. Why won't you kiss me?"

The man said, "Look, I'm a software engineer. I don't have time for a girlfriend, but to have a talking frog is cool."

Project

Communicate with others!
Step 1: Within each group, discuss how to make a story book with desktop publishing software.
Step 2: Within each group, co-work to write a short story in English.
Step 3: Use desktop publishing software to make an English story book.

Self-checklist

根据实际情况，从A、B、C、D中选择合适的答案：A代表你能很好地完成该任务；B代表你基本上可以完成该任务；C代表你完成该任务有困难；D代表你不能完成该任务。

A B C D

☐ ☐ ☐ ☐ 1. 能掌握并能运用本单元所学重点句型、词汇和短语。

☐ ☐ ☐ ☐ 2. 能理解并正确模仿听说部分的句子，正确掌握发音及语调。

☐ ☐ ☐ ☐ 3. 能模仿句型进行简单的对话。

☐ ☐ ☐ ☐ 4. 能读懂本课的短文，并正确回答相关问题。

☐ ☐ ☐ ☐ 5. 能掌握方式、结果、地点、原因状语的用法。

☐ ☐ ☐ ☐ 6. 能掌握方式、结果、地点、原因状语的引导词的用法。

☐ ☐ ☐ ☐ 7. 能掌握前缀re-的派生词的构词方法。

☐ ☐ ☐ ☐ 8. 能用课文中学习的词组造句。

☐ ☐ ☐ ☐ 9. 能向同学介绍桌面排版软件等相关知识。

Unit 10 Computer Use

Unit Goals

In this unit, you will be able to
- understand listening materials about computer use;
- talk about computers use;
- understand articles about computer use in daily life;
- write a short paragraph about computer use;
- demonstrate your knowledge about computer use.

Lead-in

Picture matching.

1. surf the internet
2. chat online
3. play computer games
4. send e-mails

Listening and Speaking

Listen and complete.

a. What else do you do with computers?
b. I often surf the Internet to get some useful information for my assignment.
c. So you cannot pay with your credit card if the computer is down.
d. Could you spare me a few minutes?
e. I'm a net surfer and a music lover.
f. The bar codes are scanned and read by the computer so that we will know whether the book is available or not.

Do you often use computers?

Amy, a journalist, is going to write a report about the use of computers in our work and life. She is interviewing some people.

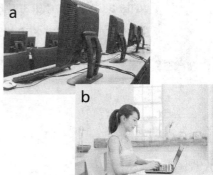

Amy is now outside the city library, interviewing Helen, a college student.

❶   Amy: Excuse me. Do you often use computers at school?

Helen: Yup. _____ I write my papers with the computer, too, because my professor asks us to have our papers printed out.

Amy: Do you enjoy using computers in your daily life?

Helen: Of course. Actually I cannot imagine a life without computers. It is an indispensable tool in this Information Age.

Amy turns to Peter, a boy student.

2 Amy: Do you use computers quite often?

Peter: Definitely. _____ I'm fond of music. Look at my iPhone, the birthday gift my parents bought me. Isn't it smart? Each week I download some music from the iTunes Store.

Amy: It's really smart. _____

Peter: I download some antivirus software. Sometimes i also play games with others online. It's great fun. I really enjoy that.

Amy: Oh, I see. OK. Thanks for your time.

Amy turns to a woman in her early fifties, and does an interview of her.

3 Amy: Excuse me, madam. _____ I'm a journalist from the local newspaper.

Madam: OK, if it's not too long.

Amy: Do you use computers at work?

Madam: Yeah, a lot. I'm a librarian at this library. All our books and reader information are stored in the computer database. We use computers to search and retrieve information.

Amy: I notice that on the back of books there are bar codes. Why?

Madam: The bar codes are very important. _____

Jack, a passer-by, chimes in.

❹ Jack: I also use computers quite often.
Amy: Where do you work, sir?
Jack: Well, I work in a bank. We use computers to keep and record the accounts of our customers. _____
Amy: It's really nice to have a talk with you. Thank you for your time.

Role Play

A: Do you enjoy using computers in your daily life?
B: Yes, very much.
A: What do you usually do with it?
B: Well, I usually use it to write my papers / download … / search and retrieve information …

Words & Expressions

credit /ˈkredɪt/ *n.* 信用
surfer /ˈsɜːfə(r)/ *n.* 冲浪者
scan /skæn/ *v.* 扫描
chime /tʃaɪm/ *v.* 插嘴
antivirus /ˈæntivaɪrəs/ *adj.* 抗病毒
librarian /laɪˈbreəriən/ *n.* 图书管理员

passer-by /ˌpɑːsəˈbaɪ/ *n.* (pl. passers-by) 过路人
bar code 条形码
chime in 插话

Unit 10 Computer Use

Reading

Text A

Pre-reading activities.

1. Discuss the advantages and disadvantages of computers with your friends. Report your findings in the class.

2. Find the following signal words in the text.

| yet | therefore | nevertheless |
| moreover | furthermore | on the other hand |

Reading Strategy

篇章信号词 Signal Words 在阅读中，篇章信号词会提示读者将在下一个句子中读到的内容。所以理解信号词有助于弄清楚句子之间的关系，更好地理解篇章内容，比如，therefore 表示句子间的因果关系，nevertheless、yet 表示转折关系，moreover、furthermore 表示递进关系，on the other hand 表示对比关系等。

Advantages and Disadvantages of Computer Use

Computers are time-saving devices that make our work easier and more effective. Computer technology creates the Internet, which has become an important source of information. Moreover, it enables students to increase their knowledge and develop their research skills. Nevertheless, computers are addictive. It is easy to lose contact with the real life by playing computer games, chatting in the chat rooms or surfing the Internet for long hours. Furthermore, spending too much time in front of a computer screen results in weakness of sight or pain in the back.

Computer technology enables us to do shopping, pay bills, or contact our friends quickly and easily. On the other hand, the number of crimes committed by hackers is increasing. Therefore, there is a danger that somebody will hack into your computer system and steal some important data.

Computer technology is indispensable to the future development of our civilisation. Yet, we must remember that computers are only machines, which cannot replace human beings. However useful they might be, they are here for us, not the other way round.

Words & Expressions

nevertheless /ˌnevəðəˈles/ *adv.* 然而
moreover /mɔːrˈəʊvə(r)/ *adv.* 而且
furthermore /ˌfɜːðəˈmɔː(r)/ *adv.* 而且，此外
disadvantage /ˌdɪsədˈvɑːntɪdʒ/ *n.* 劣势，短处
addictive /əˈdɪktɪv/ *adj.* 易上瘾的，使人入迷的
crime /kraɪm/ *n.* 罪恶
commit /kəˈmɪt/ *v.* 犯(错误)，做(坏事)

hacker /ˈhækə(r)/ *n.* 黑客
hack /hæk/ *v.* 劈，砍；破坏
civilisation /ˌsɪvəlaɪˈzeɪʃn/ *n.* 文明

enable sb to do sth 使某人能做某事
lose contact with 和……失去联系
in front of 在……前面
result in 导致
on the other hand 另一方面
the other way round 从相反方向

I. List the advantages and disadvantages of computers according to the text.

Advantages	Disadvantages

II. Connect the two sentences with an appropriate signal word.
1. I have a computer. I can work on it. (therefore)
2. Computers can store data in the form of pictures. Computers can store data in the forms of audio and video. (moreover)
3. Sometimes, inappropriate information is accessed. Sometimes, inappropriate information is used for crimes like hacking, stealing personal information, etc. (furthermore)
4. Computers have advantages. Computers have disadvantages. (on the other hand)

Text B

Pre-reading questions.
1. What role do computers play in our lives?
2. In which fields are computers used?

Computers' Effects Upon Our Lives

In this digital age, computers have affected our lives in many ways —

from home to school to the workplace. They are widely used in all walks of life. Banks, shops, offices, airports and classrooms are just a few of them. They have already revolutionised our lives more than anything that we have invented so far and today it is hardly possible to imagine a world without them.

At schools computers are used in classroom teaching and distance learning. Many students now can receive education of high quality without travelling long distances. And they will find lessons more vivid and fun when their teachers deliver lectures in smart classrooms.

Moreover, computers are used in many other ways. Computers can be put inside some household appliances, such as mobile phones, microwave ovens, digital cameras, light pens, washing machines and so on. A TV's remote control is another example. It helps people who cannot do their tasks manually.

As technology develops, computers will continue to change our lives in many more amazing ways.

Words & Expressions

workplace /'wɜːkpleɪs/ n. 工作场所，车间
revolutionise /ˌrevə'luːʃənaɪz/ v. 改革；使完全不同
distance /'dɪstəns/ n. 距离，远程
education /ˌedʒu'keɪʃn/ n. 教育
vivid /'vɪvɪd/ adj. 生动的
deliver /dɪ'lɪvə(r)/ v. 陈述，发言；递送
lecture /'lektʃə(r)/ n. 演讲，讲课

household /'haʊshəʊld/ adj. 家用的，家庭的
appliance /ə'plaɪəns/ n. 器具，器械
remote /rɪ'məʊt/ adj. 遥远的

so far 到目前为止
walk/walks of life 行业；职业
smart classroom 智慧教室

Unit 10 Computer Use

True or false.

1. Computers are only used in a few public places.
2. Computers have greatly changed our lives.
3. Students now have to travel long distances to receive education.
4. Computer technology is widely applied to household machines.
5. We are not sure whether computers will change our lives in the future or not.

Notes

smart classroom 智慧教室：智慧教室是一种典型的智慧学习环境的物化，是多媒体和网络教室的高端形态，它是借助物联网技术、云计算技术和智能技术等构建起来的新型教室，该新型教室包括有形的物理空间和无形的数字空间，通过各类智能装备辅助教学内容呈现、便利学习资源获取、促进课堂交互开展，实现情境感知和环境管理功能。智慧教室旨在为教学活动提供人性化、智能化的互动空间；通过物理空间与数字空间的结合，本地与远程的结合，改善人与学习环境的关系，在学习空间实现人与环境的自然交互，促进个性化学习、开放式学习和泛在学习。

distance learning 远程教育：又称远距教学、遥距教育，是指使用电视及互联网等传播媒体的教学模式，它突破了时空的界线，有别于传统的学校课堂教学模式，其学生通常是业余进修者。学生不需要到特定地点，因此可以随时随地上课。学生亦可以通过电视广播、互联网、辅导专线、面授(函授)等多种不同方式自助学习。

计算机的应用领域

　　计算机的应用领域主要有：
　　1. 科学计算(或数值计算)
　　2. 数据处理(或信息处理)

3. 辅助技术(或计算机辅助设计与制造)
4. 过程控制(或实时控制)
5. 人工智能(或智能模拟)
6. 网络应用
7. 语言翻译

  随着云计算、物联网、移动计算、大数据、人工智能、区块链等计算技术的快速发展和融合，计算机将应用于和渗透到经济社会生活的各个方面。

Language Practice

I. Fill in the blanks with the missing letters of the words according to the pictures.

1. bar c_d_

2. d_l _ ver a lecture

3. a ha_k_r

4. ad_ _ _t_ve

5. computer scr_ _n

6. ch_ t

7. househ_ _d appli_ _ces

8. sm _ _ t classroom

II. Complete the sentences with the words practiced above.

1. In a supermarket, every piece of goods has a _____.
2. _____ can be dangerous because they may steal or destroy important information in computers.
3. A _____ is used to display the output of a computer to the user.
4. Professor Smith is going to _____ on the latest development of the Internet technology at 10:00 tomorrow in Room 102.
5. In this school, students have lessons in _____.
6. People can _____ with each other via the Internet any time and anywhere.
7. Some children are _____ to computer games.
8. All kinds of _____ have made our lives easier and more comfortable.

III. Complete the passage with the words in the box.

data	entertainment	communication
multiple	easier	convenient

People use computers as they make their jobs _____. They can be used for _____ purposes, to store and calculate _____ and to write up massive documents _____ times while only needing to write it up once. Others use computers for _____ value, i.e. to play games, watch movies, play music, etc. Computers are also more _____ and reliable than the older ways of doing things, like typewriters, scales and other counting devices.

Ⅳ. Sentence completion.
1. Now computers are used in _____. (现在计算机在各个行业都得到了应用。)
2. _____ the government has managed to prevent the hackers' attack. (到目前为止，政府成功地防止了黑客的攻击。)
3. Sitting in front of TVs for a long time _____ weakness of sight or pain in the back. (长时间坐在电视前会引起视力减弱、后背酸痛。)
4. Henry has _____ his computer teacher. (亨利和他的计算机老师失去了联系。)
5. Computers _____ us _____ work more easily and live a much better life. (计算机使我们的工作更便捷，生活更美好。)

Writing

Write a short paragraph (at least 5 sentences) about computer use in our daily life.

Grammar

Attributive Clause (定语从句) (1)		
person	who	They help people **who** cannot do their tasks manually to automate them.
	whom	The software engineer **whom** Peter met showed him how to operate the machine.

thing	which	Many students now can receive education **which** is of high quality without travelling long distances.
person or thing	that	Computers have already changed our lives more than anything **that** we have invented so far.
	whose	She lives in the house **whose** door and windows are broken.

I. Multiple choices.

1. Here is Mr Smith _____ is going to teach us computer science this term.
   A. whose   B. who   C. which
2. Inside a computer there is a circuit board _____ is called motherboard.
   A. which   B. who   C. whom
3. Who was the scientist _____ first invented computer?
   A. whose   B. which   C. that
4. Bill Gates is a great programmer _____ name is known to many people.
   A. that   B. whose   C. which
5. There are two kinds of memory _____ are called ROM and RAM.
   A. whom   B. who   C. which

II. Connect the two sentences with a relative pronoun.

1. A young man is working on a computer. He is a software engineer. (who)
2. The boy asked a question. The question is about computer use. (which)
3. The student often plays computer games. His father works in Microsoft. (whose)
4. Computers help people automate various tasks. People cannot do the tasks manually. (that)
5. Tony chatted with a girl in the chat room. The girl is a dancer. (whom)

Word formation — Derivation (派生法)

前缀en- / em- (用于b, m, p开头的单词前) 置于名词、形容词前构成动词，表示"使成为、使成某种状态"，如enable；micro-置于名词、形容词前，表示"微、小"，如：microwave；auto-置于名词、动词、形容词前，表示"自己的、由自己的、独立的"，如automate等。

III. Add prefixes en- / em-, auto- or micro- to the following words to form derivative words and tell their Chinese meanings.

en / em+noun / adjective	verb	micro+noun / adjective	noun / adjective	auto+noun / verb / adjective	noun / verb
power		computer		mate	
courage		scopic		dial	
rich		processor		mobile	

Game

Across

1. Nowadays smart phones have become necessary _____s in our daily life.
2. Sorry, you cannot pay with your _____ card.
3. Lessons become more _____ and fun when teachers deliver lectures in smart classrooms.
4. Many students now can receive _____ of high quality without travelling long distances.

5. Professor Li is going to _____ a lecture online tomorrow afternoon.
6. With the use of smart classrooms, professors' _____s become more attractive.

Down

1. _____s are those who commit crimes online.
2. Computers are _____, so don't allow kids to play on it for long hours.
3. Every advantage has its _____.
4. Do you think you can escape the punishment after you _____ such a bad thing?

Fun Time

Humour

It was in a computer store one day. I heard a woman say to the salesman, "I want a game. It should be able to hold the interest of my six-year-old son, and simple enough for his father to play, too."

Project

Communicate with others!

Step 1: Within each group, make a list of common problems or issues in computer use.
Step 2: Have a discussion in each group about solutions to the problems.
Step 3: Invite some students to report the results.

Self-checklist

根据实际情况，从A、B、C、D中选择合适的答案：A代表你能很好地完成该任务；B代表你基本上可以完成该任务；C代表你完成该任务有困难；D代表你不能完成该任务。

A B C D

☐ ☐ ☐ ☐ 1. 能掌握并能运用本单元所学重点句型、词汇和短语。

☐ ☐ ☐ ☐ 2. 能理解并正确模仿听说部分的句子，正确掌握发音及语调。

☐ ☐ ☐ ☐ 3. 能模仿句型进行简单的对话。

☐ ☐ ☐ ☐ 4. 能读懂本课的短文，并正确回答相关问题。

☐ ☐ ☐ ☐ 5. 能掌握以who，which，that，whose，whom引导的定语从句的用法。

☐ ☐ ☐ ☐ 6. 能掌握以en- / em-，micro-及auto-为前缀的派生词的构词方法。

☐ ☐ ☐ ☐ 7. 能用课文中学习的词组造句。

☐ ☐ ☐ ☐ 8. 能向同学介绍关于计算机用途的相关知识。

Unit 11

Network

Unit Goals

In this unit, you will be able to
- understand listening materials about network;
- talk about network;
- understand articles about an introduction to computer network;
- write a short paragraph about computer network;
- demonstrate your knowledge about network.

Lead-in

Picture matching.

1. bus network
2. ring network
3. star network
4. optical fibre
5. video conferencing

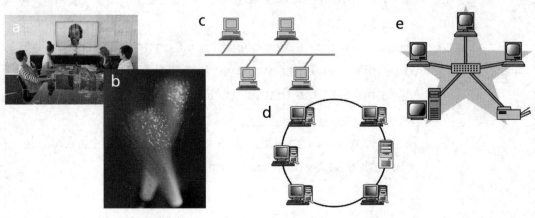

Listening and Speaking

Listen and complete.

a. I can play computer games against my little brother.
b. Our school network allows us to access the school library.
c. They share data although they are in different cities all through the country.
d. My uncle runs a dotcom business.

Where Are Networks Used?

Diana, a computer teacher, is holding a discussion among her students about where networks are used.

Tim stands up first and presents his idea.

Tim: I think networks exist extensively in our daily life.
Diana: Can you give us an example?
Tim: A good example is banks.
Diana: I know your father works in China Construction Bank. Can you explain how a network works in banks?
Tim: My father tells me that all its branches are connected via a network. _____
Diana: That's why you can draw money at the school bank while your father saves money at a bank home.

Unit 11 Network

The whole class burst into laughter and Diana turns to Paula.

2   Diana: Paula, can you give us an example that networks exist around us?

       Paula: _____ He has 12 workers and 5 offices.

       Diana: Does he also have a network?

       Paula: The answer is yes. They can share files and use the same printer.

Kim cannot wait and starts to talk.

3   Kim: I think our school is networked.

       Diana: Excellent. Then how is your school life affected by our school network?

       Kim: _____

       Diana: And what else can a school network do for you?

       Kim: Mm, I can hand in my homework through network.

       Diana: That's true.

At this moment, the naughty Andy chimes in.

       Andy: I hope my house is mini-networked.

4   Diana: But what for?

       Andy: _____ All I need is Wi-Fi and some cheap software.

       Diana: Yes. Networks can do a lot for us. If you work hard on computers you can make networks work better for us.

Role Play

A: Where are computer networks used?

B: Well, they are used at schools / in businesses / in banks / at work ...

Words & Expressions

ring /rɪŋ/ *n.* 环形物；环状
fibre /ˈfaɪbə(r)/ *n.* 光纤；纤维
extensively /ɪkˈstensɪvli/ *adv.* 广泛地
construction /kənˈstrʌkʃn/ *n.* 建设
branch /brɑːntʃ/ *n.* 枝，分枝；分部
via /ˈvaɪə, ˈviːə/ *prep.* 通过
naughty /ˈnɔːti/ *adj.* 顽皮的，淘气的

video conferencing 视频会议
burst into laughter 突然大笑
turn to 转向
to one's surprise 出乎……的意料
hand in 上交

Reading

Text A

Pre-reading questions.

1. What do you know about computer networks?

2. Why do people use networks?

3. When you read a long sentence, which part or parts of the sentence would you try to understand first? Why?

Reading Strategy

分析长难句结构 Sentence Structure Analysis 英语长难句里面可能会有多个从句，从句与从句之间的关系可能为并列、包含与被包含、镶嵌等，而两个本应紧密相连的句子成分也可能被其他成分分开。有些词有前置或后置的不定式、分词等形式的修饰成分，给阅读造成困难。分析长难句，首先要找出全句的主语、谓语和宾语，即句子的主干结构，其次找出句中所有的谓语结构、非谓语结构、介词短语和从句的引导词，然后分析从句和短语的功能，

例如，是否为主语从句、宾语从句、表语从句或状语从句等，以及词、短语和从句之间的关系，分析句子中是否有固定词组或固定搭配、插入语等其他成分。

Computer Networks

A computer network, often simply referred to as a network, is a collection of computers and devices interconnected by communications channels that facilitate communications among users and allow users to share resources.

Networks can be classified according to the hardware and software technology that is used to interconnect the individual devices in the network, such as optical fibre, LAN, Ethernet, WLAN, etc. They may also be classified according to the network topology upon which the network is based, such as bus network, star network, ring network, etc.

Networks can be used for a variety of purposes. Using a network, people can communicate efficiently and easily via emails, chat rooms, video telephone calls, and video conferencing. In a networked environment, each computer may access and use hardware resources on the network, such as printing a document on a shared network printer. Authorised users may access data and information stored in other computers on the network. Many networks provide access to data and information on shared storage devices. Users connected to a network may run application programs on remote computers.

Words & Expressions

collection /kəˈlekʃn/ *n.* 一批物品

channel /ˈtʃænl/ *n.* 通道；渠道，途径；频道

facilitate /fəˈsɪlɪteɪt/ *v.* 使便利；促进；为他人提供方便(或机会)

classify /ˈklæsɪfaɪ/ *v.* 分类

LAN /læn/ 网络，局域网，本地网

Ethernet /ˈiːθənet/ 以太网，采用共享总线型传输媒体方式的局域网

topology /təˈpɒlədʒi/ *n.* 拓扑 (网络布局结构)

base /beɪs/ *v.* 以……为基础

variety /vəˈraɪəti/ *n.* 各种

authorised /ˈɔːθəraɪzd/ *adj.* 权威认可的，经授权的

according to 按照，根据

base on / upon 以……为基础

a variety of 多种的，各种各样的

Match the phrases in the left column to the sentences in the right column with the same meaning.

1. facilitating communications	a. In a networked environment, each computer on a network may access and use hardware resources on the network, such as printing a document on a shared network printer.
2. sharing software	b. Users connected to a network may run application programs on remote computers.
3. sharing files, data, and information	c. Using a network, people can communicate efficiently and easily via email, instant messaging, chat rooms, telephone, video telephone calls, and video conferencing.
4. sharing hardware	d. In a networked environment, authorised user may access data and information stored on other computers on the network.

Unit 11 Network

Text B

Pre-reading questions.

1. How important are networks? Discuss with your partner about the importance of networks.
2. Do you know any ways that networks are linked? List the ways you know.

How Are Network Components Connected?

Networks link in several different ways. Generally speaking, a network can be linked in three ways.

Workstations, servers and other machines can be connected to a network using a single cable, and this is the reason why it is the simplest way. It is called a bus network. There are times when this is done with spurs. This linking way is easy and cheap. But it is slow and breaks down easily.

In places where the workstations are arranged in a ring and linked via a cable, a ring network is set up. A ring network is fast because all the data travel in the same direction. But if one workstation fails, the whole network goes down.

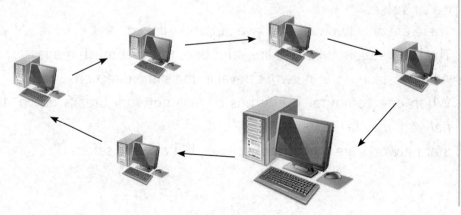

The safest one is a star network where each workstation has its own direct line to the server. If one link fails, which sometimes happens, it does not affect others. Even though it is expensive to install, a star network is normally preferred because of its safety.

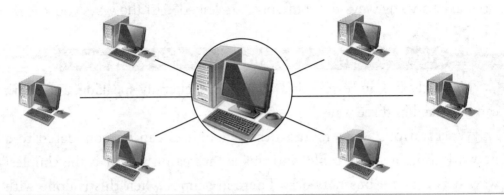

Words & Expressions

spur /spɜː(r)/ *n.* 支线
direction /dɪˈrekʃn, daɪˈrekʃn/ *n.* 方向
normally /ˈnɔːməli/ *adv.* 正常地；通常地
prefer /prɪˈfɜː(r)/ *v.* 更喜欢；宁愿

generally speaking 一般说来，通常
break down 毁掉；停顿；中止
because of 因为

True or false.
1. There are only two ways to have networks linked.
2. The bus network is easy to link, and does not cost much money.
3. All the data in a star network travel in the same direction.
4. When one computer in the bus or ring network breaks down, the whole network may fail to work.
5. Star networks are the most expensive, but also the safest.

Unit 11 Network

Notes

server 服务器：服务器是计算机的一种，它比普通计算机运行更快、负载更高、价格更贵。服务器在网络中为其它客户机（如PC机、智能手机、ATM等终端甚至是火车系统等大型设备）提供计算或者应用服务。服务器具有高速的CPU运算能力、长时间的可靠运行、强大的I/O外部数据吞吐能力以及更好的扩展性。一般来说，服务器应具备承担响应服务请求、承担服务、保障服务的能力。

LAN 局域网：Local Area Network，是在一个局部的地理范围内(如一个学校、工厂或机关内)，将各种计算机、外部设备和数据库等互相联接起来组成的计算机通信网，简称LAN。它可以通过数据通信网或专用数据电路，与远方的局域网、数据库或处理中心相连接，构成一个大范围的信息处理系统。

a bus network 总线网络：总线网络是连接工作站、服务器以及其他计算机最简单的方式。它们之间只通过一根电缆连接。有时，总线网络也会有一些分支。这种连接方式简便、造价低，但速度慢、易死机。

a ring network 环形网络：此种网络由电缆连接成环状。它的速度快，因为所有的数据都沿同一方向传输，但是一旦其中某个工作站出现问题，整个网络就会瘫痪。

a star network 星形网络：星形网络是最安全的，因为每个工作站都是与服务器单独连接，某个连接出了问题并不影响其他连接。虽然它造价高，但因其安全性高而受到普遍青睐。

Do You Know

服务器

服务器指一个管理资源并为用户提供服务的计算机软件，通常分为文件服务器、数据库服务器和应用程序服务器。运行以上软件的计算机或计算机系统也被称为服务器。相对于普通PC来说，服务器在稳定性、安全性、性能等方面都要求更高，因此CPU、芯片组、内存、磁盘系统、网络等硬件和普通PC有所不同。

局域网 (LAN)

　　局域网是由一组计算机及相关设备通过共用的通信线路（电缆）或无线连接的方式组合在一起的系统。局域网可以实现文件管理、应用软件共享、打印机共享、工作组内的日程安排、电子邮件和传真通信服务等功能。局域网是封闭型的，可以由办公室内的两台计算机组成，也可以由一个公司内的上千台计算机组成。

无线局域网(WLAN)

　　无线局域网指在不采用传统电缆的同时，提供传统有线局域网的所有功能。有两种接入方式：Wi-Fi接入方式、移动接入方式。

客户端(Client)

　　客户端又称为用户端，是指与服务器相对应，为客户提供本地服务的程序。

城域网(MAN)

　　城域网是在一个城市范围内所建立的计算机通信网，所采用的技术基本上与局域网相类似，只是规模上要大一些。城域网既可以覆盖相距不远的几栋办公楼，也可以覆盖一个城市；既可以是私人网，也可以是公用网。

广域网 (WAN)

　　广域网是一种用来实现不同地区的局域网或城域网的互连，可提供不同地区、城市和国家之间的计算机通信的远程计算机网，通常跨接很大的物理范围，所覆盖的范围从几十公里到几千公里，它能连接多个城市或国家，或横跨几个洲，并能提供远距离通信，形成国际性的远程网络。

Unit 11 Network

Language Practice

I. Fill in the blanks with the missing letters of the words according to the pictures.

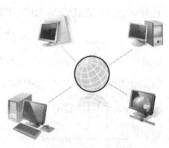

1. vid_ _ phone call

2. op_ _cal f_ber

3. a computer n_tw_ _k

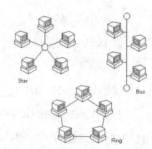

4. pr_ _t the document

5. t_po_ogy

6. con_ _ct

7. sp_rs

8. li_k

177

II. Complete the sentences with the words practiced above.

1. If you _____ your PC to the network, you can share the data and information.
2. Henry gave his son a _____ when he was in Nanjing on business.
3. Television stations around the world are _____ by satellite.
4. An _____ is a thin, transparent fibre that acts as a "light pipe" to transmit (传输) light between the two ends of the fibre.
5. Sometimes a bus network is done with _____.
6. Sally went to the business center to _____ that her boss needed.
7. With _____, we can work more easily.
8. A network _____ is the pattern in which nodes (节点) are connected to a local area network or other network via links.

III. Complete the passage with the words in the box.

| information | wireless | include |
| communication | devices | computers |

A personal area network (PAN) is used for _____ among computers and different _____ technological devices close to one person. Some examples of devices that are used in a PAN are personal _____, printers, fax machines, telephones, etc. A PAN may _____ wired and wireless _____. A wired PAN is usually constructed with USB and Firewire (槽口，火线接口) connections while technologies such as Wi-Fi, Bluetooth and infrared (红外线的) communication typically form a _____ PAN.

IV. Sentence completion.

1. Networks can be classified _____ the hardware and software technology. (网络可以根据软件和硬件技术来分类。)
2. Networks can be used for _____. (网络能够用于各种各样的用途。)

Unit 11 Network

3. _____, a network can be linked in three ways. (一般说来，网络有三种连接方式。)

4. A star network is usually preferred _____ safety. (出于安全考虑，星形网络往往成为首选。)

5. A star network is the safest because if one link _____, it will not affect others. (星形网络最安全，因为一条连接断了，不会影响其他的连接。)

Writing

Write a short paragraph (at least 5 sentences) about computer networks. You may write it by answering the following questions.

What is a computer network?
How is it classified?
What are the purposes of networks?
What is the safest way to connect network components?

Grammar

		Attributive Clause (定语从句) (2)
time	when	There are times **when** this is done with spurs.
place	where	In places **where** the workstations are arranged in a ring and linked via a cable, a ring network is set up.
reason	why	Workstations, servers and other machines can be connected to a network using a single cable, and this is the reason **why** it is the simplest way.
Restrictive Attributive Clause		The safest one is a star network **where** each workstation has its own direct line to the server.
Non-Restrictive Attributive Clause		If one link fails, **which** sometimes happens, it does not affect others.

179

I. Multiple choices.

1. I still remember the day _____ I first bought a computer.
   A. where			B. why			C. when
2. Sarah went to visit the building _____ there is a LAN (local area network).
   A. when			B. where		C. why
3. In a ring network, all the data travel in the same direction, and this is the reason _____ it is fast.
   A. why			B. where		C. when
4. Those _____ are good at computers should know the skills of typing or keyboarding.
   A. whose			B. who			C. whom
5. The software problem, _____ is very complex, has been solved.
   A. which			B. that			C. where

II. Fill in the blanks with the correct words in the box.

why	when	which	where

Different kinds of networks can be formed _____ different types of channels are used, such as cable or Wi-Fi. Telephone lines may connect communication equipment within the same building _____ it is possible to do so. Anyone can create his own network, _____ he may be glad to do, in his home. Networks may also be citywide and even international. Three important networks are LANs, MANs, and WANs according to their geographical sizes. The reasons _____ people choose one network instead of the others vary, but generally speaking they do so for their own conveniences.

Unit 11 Network

Word formation — Derivation (派生法)

后缀-age放在动词后，使其变为名词，表示行为、结果或状态，如：storage等。后缀-(i)ty放在形容词后，使其变为名词，表示性质或状态，如：safety, variety等。

III. Add –age or –(i)ty to the following words to form derivative words and then tell their Chinese meanings. Make changes where necessary.

verb+-age	noun	adjective+-(i)ty	noun
store		safe	
cover		various	
block		special	

Game

Across

1. Optical _____ has been applied to China's telecommunications for quite some time.
2. The third stage of data _____ involves putting the data in order.
3. _____ technology includes GPS units, wireless computer mouses, keyboards, satellite television and cordless telephones (无绳电话).
4. What kind of network is _____ed in your school?

Down

1. Before I send the email, I have to _____ the computer to the Internet.
2. The school computer network has greatly _____d the communication among students.
3. The computer did not have its original _____ parts.
4. A ring network is set up when the _____s are arranged in a ring.
5. We know that a ring network is fast because all the data travel in the same _____.
6. I _____ Windows 10 to 7, for it is more stable.

Fun Time

My kids loved surfing the web, and they wrote their passwords on Post-it notes. I noticed their Disney password was "MickeyMinnieGoofyPluto", so I asked why it was so long.

"Because," my son explained, "they say it has to have at least four characters."

Project

Communicate with others!

Step 1: Within each group, talk about where networks are used in people's life and work.

Step 2: Share with your group members the networks you often use.

Step 3: Ask one classmate from each group to give an oral presentation to summarise the networks his or her group members often use.

Unit 11 Network

Self-checklist

根据实际情况，从A、B、C、D中选择合适的答案：A代表你能很好地完成该任务；B代表你基本上可以完成该任务；C代表你完成该任务有困难；D代表你不能完成该任务。

A B C D

☐ ☐ ☐ ☐ 1. 能掌握并能运用本单元所学重点句型、词汇和短语。

☐ ☐ ☐ ☐ 2. 能理解并正确模仿听说部分的句子，正确掌握发音及语调。

☐ ☐ ☐ ☐ 3. 能模仿句型进行简单的对话。

☐ ☐ ☐ ☐ 4. 能读懂本课的短文，并正确回答相关问题。

☐ ☐ ☐ ☐ 5. 能掌握where，when和why引导的定语从句的用法。

☐ ☐ ☐ ☐ 6. 能掌握限定性和非限定性定语从句的用法。

☐ ☐ ☐ ☐ 7. 能掌握以-age和-(i)ty为后缀的派生词的构词方法。

☐ ☐ ☐ ☐ 8. 能用课文中学习的词和词组造句。

☐ ☐ ☐ ☐ 9. 能向同学介绍电脑网络及其种类等相关基本知识。

Unit 12

The Internet

Unit Goals

In this unit, you will be able to
- understand the Internet information;
- talk about the Internet;
- understand articles about the Internet;
- write a short paragraph about what you can do with the Internet;
- demonstrate your knowledge about the Internet.

 Lead-in

Picture matching.

1. browser
2. chat room
3. news group
4. online advertising
5. e-commerce
6. web utilities
7. e-learning site
8. online shopping
9. catalogue shopping

Unit 12 The Internet

Listening and Speaking

Listen and complete.

a. Let's see if there is anything new about basketball.
b. You can chat online with some silver netters using QQ or WeChat.
c. I hope you can sign up for a course online to improve your French.
d. Of course you can see that person if he has a camera.
e. It's more convenient than catalogue shopping.

What Can I Do with the Internet?

Johnny's family has just got the Internet installed at home. They are very exited and talking about what they will do with the Internet.

❶ Johnny: Great! I can now access the Internet at home.
　　Mom: Johnny, you cannot play games online every day.
　　Johnny: Sure, mom. But I will be in touch with my friends via email.
　　Mom: _____

❷ Mom: What can I do with the Internet?
　　Dad: You can shop online. _____
　　Mom: Really? Then how should I pay?
　　Dad: By e-banking. They usually charge with credit cards.

185

Mom: Is it safe? I doubt it.

3 Dad: I can now get the latest sports news online. _____
Mom: I have no interest in sports. I want to read about the latest fashions.
Dad: Well, I've got a suggestion. How about a movie online?
Mom: A great idea.

4 Grandpa: The Internet is for young people. It's of no use for me.
Johnny: Grandpa, _____
Grandpa: But I don't know how to type.
Johnny: You don't have to type. You can talk via microphone.
Grandpa: It's no fun to talk to the screen.
Johnny: _____
Grandpa: It sounds fantastic.

Role Play

A: What do you usually do with the Internet?
B: Well, I usually use it to download music / chat with my friends / play games / ...

Words & Expressions

advertise /ˈædvətaɪz/ v. 做广告，登广告
e-commerce /iːˈkɒmɜːs/ n. 电子商务
catalogue /ˈkætəlɒg/ n. 目录；目录册
silver /ˈsɪlvə(r)/ adj. 银发的
netter /ˈnetə/ n. 上网冲浪的人
messenger /ˈmesɪndʒə(r)/ n. 报信者；邮递员
e-bank /ˈiːbæŋk/ n. 电子银行
doubt /daʊt/ v. 怀疑

sign up for 报名参加
be in touch with 与……联系

silver netter 上年纪的上网者

Reading

Text A

Pre-reading activities.

1. Do you surf the Internet frequently?

2. What does WWW stand for, World Wide Wait or World Wide Web?

Reading Strategy

列提纲 (Outlining) 指学习者用条文式的简洁字句对学习材料进行的言语归纳。它可以说是一篇作品的骨架。列提纲便于抓住要领，理清思路，及时把文章结构有条理地显示出来。

The Internet

The Internet is an international network of computers. It was developed in 1969 in the United States. The Web, which is also known as WWW or World Wide Web, was introduced in 1992 in Switzerland.

Before the Web, the Internet was all text — no pictures, sound or videos. The Web provided a multimedia interface to resources available on the Internet. Both of them are now important tools for us to use.

It is easy to get the Internet and the Web confused, but they are not the same thing. The Internet is the actual network, made up of devices and protocols. It links computers and resources all around the world. The Web only refers to the interface to the resources.

Each day millions of people use the Internet and the Web. The most common applications of them are in communication, shopping, searching, entertainment and education.

Words & Expressions

multimedia /ˈmʌltiˈmiːdiə/ *n.* 多媒体；多媒体的采用

interface /ˈɪntəfeɪs/ *n.* 界面

confused /kənˈfjuːzd/ *adj.* 混淆的；困惑的

actual /ˈæktʃuəl/ *adj.* 实际的

protocol /ˈprəʊtəkɒl/ *n.* （数据传输的）协议

make up of 由……构成，组成

I. Complete the following outlines according to Text A.
1. It's a brief introduction of the _____ and the _____.
2. The Web provided a multimedia _____.
3. It tells about the _____ between the Internet and the Web.
4. It introduces the most common _____ of the Internet and the Web.

II. True or false.
1. The Internet is also called World Wide Web.
2. The Internet started in the United States in 1969.
3. World Wide Web is a multimedia interface.
4. World Wide Web is made up of devices and protocols.
5. The Internet and World Wide Web have made e-learning possible.

Unit 12 The Internet

Text B

Pre-reading questions.
1. Discuss your use of e-mails with your partner.
2. Do you practice your English with your e-pals?

> **Email**
>
> Email is the shortened name for "electronic mail". It is one of the most common Internet activities. People use it to keep in touch with their families and friends, and to do business. For example, Outlook Express is a built-in email program that comes with Internet Explorer.
>
> The main purpose of email is to send and receive email messages. You can also send photos, sound and video files as attachments. But you must have an account on a mail server, which is similar to having an address in receiving snail mail letters by post.
>
> To set up a new email account you have to provide information about yourself and choose an account name and a password. Your account name or ID becomes part of your email address. If you open a Hotmail account and choose "ilovecomputers" as your ID, your address becomes "ilovecomputers@hotmail.com".

Words & Expressions

e-pal /'iːpæl/ *n.* 网友
electronic /ɪˌlekˈtrɒnɪk/ *adj.* 电子的
mail /meɪl/ *n.* 邮件
activity /ækˈtɪvəti/ *n.* 活动
main /meɪn/ *adj.* 主要的
attachment /əˈtætʃmənt/ *n.* 附件

similar /ˈsɪmələ(r)/ *adj.* 相似的
snail /sneɪl/ *n.* 蜗牛
password /ˈpɑːswɜːd/ *n.* 密码,口令

be similar to 与……相似
snail mail 通过邮局寄的信件

Short-answer questions.

Answer the following questions according to Text B.

1. What is the full name of "email"?

   _____

2. What do people do with emails?

   _____

3. What is the main purpose of emails?

   _____

4. What else can people do with emails besides the main purpose?

   _____

5. To set up a new email account, what do you have to provide?

   _____

Notes

email 电子邮件：电子邮件是一种用电子手段提供信息交换的通信方式，是互联网应用最广的服务。通过电子邮件系统，用户可以以非常低廉的价格（只需负担网费）、非常快速的方式（几秒钟之内可以发送到世界上任何指定的目的地），与世界上任何一个角落的网络用户联系。电子邮件可以是文字、图像、声音等多种形式。

Internet 因特网：Internet是International Net的简写，又称国际互联网。它最早产生于美国国防部的高级研究规划署。最初的目的只是远程计算机的数据共享，后来它将世界各地的计算机及计算机网络相互连接起来，形成了一个无边无际的超级大网。Internet的主要服务项目有：电子邮件 (Email)、远程登录 (Telnet)、查询服务 (Finger)、文件传输 (FTP)、文档服务器 (Archive)、新闻论坛 (Usenet)、电子公告牌 (BBS)、新闻群组 (News Group)、全球网 (World Wide Web，缩写为WWW，又称万维网) 等。

Unit 12 The Internet

Do You Know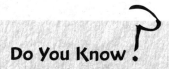

Wi-Fi

全称为wireless fidelity，在无线局域网的范畴是指"无线相容性认证"，是一种无线联网的技术，把有线网络信号转换成无线信号，以前通过网线连接电脑，而现在则是通过无线电波来连网，常见的就是一个无线路由器，在这个无线路由器的电波覆盖的有效范围都可以采用Wi-Fi连接方式进行联网。

博客

博客，又译为网络日志、部落格或部落阁等，是一种通常由个人管理、不定期粘贴新的文章的网站。博客最初的名称是Weblog，后简称为Blog。博客上的文章通常根据粘贴时间，以倒序方式由新到旧排列。许多博客是个人心中所想之事情的发表，其他博客则是一群人基于某个特定主题或共同利益领域的集体创作。一个典型的博客结合了文字、图像、其他博客或网站的链接、其他与主题相关的媒体。能够让读者以互动的方式留下意见，是许多博客的重要要素。

Language Practice

I. Fill in the blanks with the missing letters of the words according to the pictures.

1. d_v_l_p

2. ca_ _ra

3. m_ss_ _ger

191

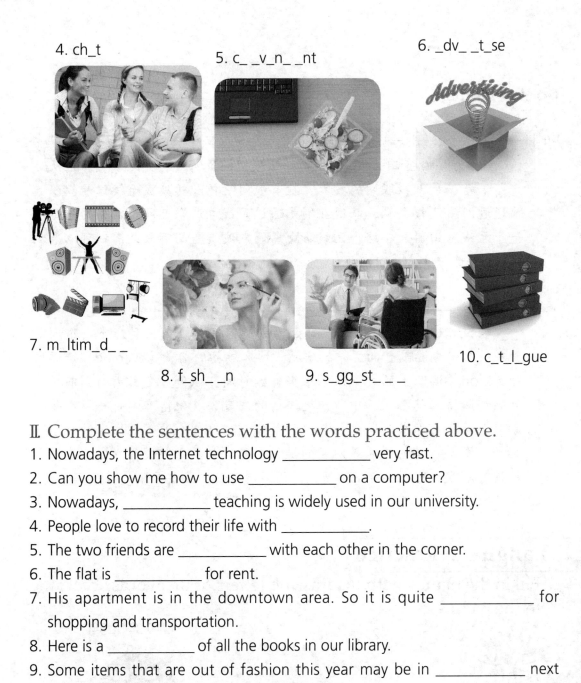

4. ch_t
5. c_ _v_n_ _nt
6. _dv_ _t_se
7. m_ltim_d_ _
8. f_sh_ _n
9. s_gg_st_ _ _
10. c_t_l_gue

II. Complete the sentences with the words practiced above.

1. Nowadays, the Internet technology _____ very fast.
2. Can you show me how to use _____ on a computer?
3. Nowadays, _____ teaching is widely used in our university.
4. People love to record their life with _____.
5. The two friends are _____ with each other in the corner.
6. The flat is _____ for rent.
7. His apartment is in the downtown area. So it is quite _____ for shopping and transportation.
8. Here is a _____ of all the books in our library.
9. Some items that are out of fashion this year may be in _____ next year.
10. Her _____ is that we should go there by air.

Unit 12 The Internet

III. Complete the passage according to the pictures.

People can access the Internet with a _____ and a _____

_____. That means accessing the Internet is possible anywhere. The cost is also getting cheaper.

In order to display web _____, software called a browser is needed. Two most popular _____ are Microsoft Internet Explorer and Google Chrome.

Many mobile phones allow limited access to the Internet, too. This system is called WAP (Wireless Application Protocol). Because of the small _____ _____ of mobile phones, only some pages can be displayed. However, it is becoming popular among young people.

193

Ⅳ. Sentence completion.

1. I hope _____ (你报名参加一个网络课程) to improve your English.
2. I will be _____ (与朋友保持联络) via WeChat.
3. The Internet is made up of _____ (设备和数据传输协议).
4. An account on a mail server _____ (与邮局寄信系统中的收信地址相似).
5. _____ (要建立一个新的电子邮件账户), you have to provide information about yourself.

Writing

Write a short paragraph (at least 5 sentences) about what we can do with the Internet.

Grammar

Object Clause (宾语从句)		
of verb	that	Anne always says **that** she has a good daughter.
	if / whether	I do not know **whether / if** he can repair the TV set.
	who	Please tell me **who** was the inventor of the telephone.
	whose	Do you know **whose** notebook is left on the table?
	what	It is common to state **what** type of data will be entered into each field.
	which ...	Please show me **which** one is yours.

Unit 12 The Internet

（续表）

	when where why how ...	I want to know **when** you have classes. They will go **where** it is quieter. I must find out **why** he is so indifferent to me. Would you tell me **how** I can get to the airport?
of preposition	what	He is rather interested in **what** the professor said.
of adjective	that	He is sure **that** they will succeed.

I. Complete the sentences with proper relative pronouns.

1. Rose told me _____ she was going to sign up for an English course online.
2. I know nothing about my new neighbor except _____ he used to work with a software company.
3. Peter asked me _____ I could keep in touch with him via WeChat.
4. We all wonder _____ Tom is so crazy about the Internet.
5. Can you tell me _____ I can get an email account?

II. Multiple choices.

1. Betty said that she _____ shopped online.
   A. has never   B. had never   C. will never
2. I hardly understand _____ Sarah has told me.
   A. what   B. which   C. that
3. You must remember _____ in his Blog.
   A. what did Tom say   B. what has Tom said C. what Tom said
4. I want to know how long _____ to download this movie.
   A. it had take me   B. it will take me   C. it took me
5. Can you tell me _____ I can get antivirus software?
   A. what   B. which   C. where

195

Word formation — Blending (混成法)

　　混成法，即将两个词混合或各取一部分紧缩而成一个新词。后半部分表示主体，前半部分表示属性，如news broadcast混成为newscast，smoke and fog混成为smog。

Ⅲ. Mix the words in column A and B to form blended words.

	A	B	Blends
1	web	magazine	
2	web	consumer	
3	web	economics	
4	net	company	
5	net	citizen	
6	screen	teenager	
7	television	computer	
8	commercial	e-mail	

Game

Across

1. Email is the shortened name for "_____ mail".
2. In the _____, you may find our family's pictures.
3. You need a _____ to enter Joe's computer.
4. Internet makes it _____ for us to get in touch with each other.
5. Online _____ing helps to display his company's products.

6. Web-based email services enable you to access email via a web _____.

Down

1. Our _____ topic today is "The Fast Development of the Internet."
2. An account on a mail server is _____ to having an address in receiving snail mail letters by post.
3. QQ is a kind of Internet _____ software.
4. The computer in our office has a very different _____.

Fun Time

He must have a computer

A mother was telling her 5-year-old son about God. "Do you know," she said to him, "that God knows where everybody is all the time, and exactly what they are doing?" The little boy looked at his mother wide-eyed and said, "Wow. He must have a computer."

Project

Communicate with others!

Step 1: Within each group, talk about the common applications of the Internet.
Step 2: Share with your group members your favourite internet application.
Step 3: Ask one classmate from each group to report in class the most common internet applications among his or her group members.

Self-checklist

根据实际情况，从A、B、C、D中选择合适的答案：A代表你能很好地完成该任务；B代表你基本上可以完成该任务；C代表你完成该任务有困难；D代表你不能完成该任务。

A B C D

☐ ☐ ☐ ☐ 1. 能掌握并能运用本单元所学重点句型、词汇和短语。

☐ ☐ ☐ ☐ 2. 能理解并正确模仿听说部分的句子，正确掌握发音及语调。

☐ ☐ ☐ ☐ 3. 能模仿句型进行简单的对话。

☐ ☐ ☐ ☐ 4. 能读懂本课的短文，并正确回答相关问题。

☐ ☐ ☐ ☐ 5. 能掌握宾语从句的种类。

☐ ☐ ☐ ☐ 6. 能掌握宾语从句的语序和关联词的用法。

☐ ☐ ☐ ☐ 7. 能掌握混成词构词法。

☐ ☐ ☐ ☐ 8. 能用课文中学习的词组造句。

☐ ☐ ☐ ☐ 9. 能向同学介绍互联网及其相关基本知识。

Unit 13
Some Important Issues

Unit Goals

In this unit, you will be able to
- understand some important computer issues;
- talk about computer issues;
- understand articles about computer issues and addiction;
- write a short paragraph about computer issues;
- demonstrate your knowledge about computer issues.

Lead-in

Picture matching.

1. privacy
2. security
3. environment
4. virus
5. computer theft
6. computer addict

a

b

c

d 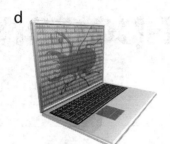 e  f

Listening and Speaking

Listen and complete.

a. I'm quite fond of computer games and I surf the Internet every day after school.
b. It's against the Intellectual Property Rights.
c. That's common for many computer addicts.
d. They may break into my system.
e. Have a drink, take a walk and stretch your muscles.
f. I've got used to being alone in the house.

A Discussion About Computer Issues

The IT teacher, Ms Garrett, is holding a discussion about some important computer issues that students have come across.

1

Alice: Computer can cause some health problems. I usually spend many hours in front of computers, so I always feel pain in my back, arms and neck.

Ms Garrett: _____ And they also frequently feel pain in eyes. More and more people are getting near-sighted.

Alice: Sounds serious. What's the cure?

200

Unit 13 Some Important Issues

Ms Garrett: Make sure you take regular breaks if you work on a computer for a long time. _____
Alice: Many thanks for your sound advice.

❷ Terry: _____ My parents are always against this.
Ms Garrett: Terry, I have to say your parents are right.
Terry: But it is the only entertainment for me. _____
Ms Garrett: That's the problem. Why not do some outdoor sports and make more friends?
Terry: Great. I will have a try.

❸ Ben: I love the convenience that computers bring about. But my only concern is the safety problem. I always fear that my computer might be attacked by a virus.
Ms Garrett: You can use antivirus software to protect your computer.
Ben: But some hackers might work out my passwords. _____
Ms Garrett: Oh, it happens rarely. But when it happens, it is a nightmare for all of us.

❹ Matthew: I am wondering if we should buy pirated software. It is really cheap.
Ms Garrett: Absolutely not. _____ It may also be of poor quality.
Matthew: I'll tell this to my friends.
Ms Garrett: Yes. We should let more people know it.

Role Play

A: What are some important computer issues?
B: Well, I think it causes some health / safety / copyright / ... problems, because ...

201

Words & Expressions

privacy /ˈprɪvəsi/ n. 隐私
security /sɪˈkjʊərəti/ n. 安全
theft /θeft/ n. 偷，行窃
addict /ˈædɪkt/ n. 对……入迷、上瘾的人
intellectual /ˌɪntəˈlektʃuəl/ adj. 知识的；智力的
 n. 知识分子
property /ˈprɒpəti/ n. 资产，财产
stretch /stretʃ/ v. 伸展

muscle /ˈmʌsl/ n. 肌肉；臂力
near-sighted /ˌnɪəˈsaɪtɪd/ adj. 近视的
issue /ˈɪʃuː/ n. 问题

intellectual property 知识产权
break into 闯入
come across 遇到
bring about 使发生，致使
work out 解决，计算出

Reading

Text A

Pre-reading activities.

1. Have you ever heard of computer viruses? List some viruses you know.

2. What can people do against computer viruses?

Reading Strategy

概述 Summarizing 概述策略指的是阅读完材料后，将所阅材料浓缩，以书面或口头的形式作出所阅读材料的梗概。概述策略通常采用归纳的方式进行，在阅读材料之后，以一定的形式(常为表格、导图等)对文中的事实部分进行罗列，以一条主线为导向，以某种逻辑为序对文章进行剖析，使文章的行文脉络清晰地呈现出来。

Unit 13 Some Important Issues

Computer Viruses

Computer users should be careful about the security of computers. One risk is that computers may be attacked or infected by viruses. The viruses either come from the Internet when you receive an email or download a program or from another computer while you copy a file.

Some people create viruses for fun or out of evil will. Viruses are computer programs, which can copy themselves. Viruses can cause terrible damage to people's computers. It takes plenty of money to fix if it happens worldwide.

Some viruses slow down the Internet by creating traffic jams. Some viruses make some or all data disappear. Some viruses cause the whole computer world to break down. Luckily we can fight against them with antivirus software. But remember to update your anti-virus program regularly.

Words & Expressions

infect /ɪnˈfekt/ v. 感染；传染
evil /ˈiːvl/ adj. 邪恶的
damage /ˈdæmɪdʒ/ n. 损害，毁坏
plenty /ˈplenti/ n. 丰富；大量
fix /fɪks/ v. 修理，修复
disappear /ˌdɪsəˈpɪə(r)/ v. 消失

luckily /ˈlʌkɪli/ adv. 幸运地
fight /faɪt/ v. 战斗；作战
regularly /ˈreɡjələli/ adv. 有规律地

plenty of 很多

I. True or false.
1. If your computer is not linked with a network, you are safe from viruses.
2. Viruses are terrible because they can copy themselves.
3. Viruses can become a worldwide problem.
4. It takes a lot of money to repair the damage caused by viruses.
5. Once you have antivirus software installed, your computer is safe forever.

II. Fill in the blanks according to Text A, and then you will get a summary.

Viruses can cause terrible damage to people's computers. They either come from the _____ or from another _____. Viruses are computer _____, which can copy themselves. Some people create them for _____ or out of evil _____. Luckily we can fight against them with _____ software.

Text B

Pre-reading questions.
1. Do you know anyone around you who is suffering from Internet addiction?
2. Do you think you are an Internet addict? Why or why not?

Internet Addiction

We all enjoy the benefits of the Internet, and for many of us it is also an important tool for work, education, and communication. While time spent on the Internet can be hugely productive, for some people compulsive Internet use can interfere with their daily lives, work and relationships.

When you feel more comfortable with your online friends than your real ones, or you cannot stop yourself from playing games or surfing, even when it has negative consequences in your life, then you may be using the Internet too much.

Many people who suffer from Internet addiction may have trouble in completing tasks at work or home, isolate from family or friends and feel guilty about their Internet use. A lot of them wouldn't have suffered so much if they had realised their overuse of the Internet earlier.

Here are some questions people can use as a check-list for overuse of the Internet:

How often do you find that you stay online longer than you intended?

How often do you form new relationships with fellow online users?

How often do others in your life complain to you about the amount of time you spend online?

How often does your study or school work suffer because of the amount of time you spend online?

Words & Expressions

addiction /əˈdɪkʃn/ n. 沉溺，上瘾
benefit /ˈbenɪfɪt/ n. 益处，好处
productive /prəˈdʌktɪv/ adj. 有成效的；生产的，多产的
compulsive /kəmˈpʌlsɪv/ adj. 上瘾的，无法控制行为的
interfere /ˌɪntəˈfɪə(r)/ v. 干预，妨碍
relationship /rɪˈleɪʃnʃɪp/ n. 关系
comfortable /ˈkʌmftəbl/ adj. 舒适的

consequence /ˈkɒnsɪkwəns/ n. 结果
suffer /ˈsʌfə(r)/ v. 受苦
isolate /ˈaɪsəleɪt/ v. 使孤立，隔离
guilty /ˈɡɪlti/ adj. 内疚的，有罪的
intend /ɪnˈtend/ v. 打算
complain /kəmˈpleɪn/ v. 抱怨

interfere with 打扰；妨碍
have trouble in 在……有困难

205

Short-answer questions.

Answer the following questions according to Text B.

1. What role does the Internet play in many people's work, education, and communication?

   _____

2. What are the negative consequences of compulsive Internet use?

   _____

3. Whom will the Internet addicts be more comfortable with, online friends or real ones?

   _____

4. What is the trouble of some Internet addicts?

   _____

5. Why do the Internet addicts feel guilty?

   _____

Notes

Internet security 网络安全：安全就是最大程度地减少数据和资源被攻击的可能性。网络安全是指网络系统的硬件、软件及其系统中的数据受到保护，不受偶然的原因或者恶意的攻击而遭到破坏、更改、泄露，系统连续、可靠、正常地运行，网络服务不中断。网络安全从其本质上来讲就是网络上的信息安全。

virus 病毒：计算机病毒是指编制或者在计算机程序中插入的破坏计算机功能或者毁坏数据、影响计算机使用，并能自我复制的一组计算机指令或者程序代码。就像生物病毒一样，计算机病毒有独特的复制能力，它可以很快地蔓

Unit 13 Some Important Issues

延，又常常难以根除。它们能把自身附着在各种类型的文件上，当文件被复制或从一个用户传送到另一个用户时，它们就随同文件一起蔓延开来。

Intellectual Property Rights 知识产权：知识产权是指自然人或法人对自然人通过智力劳动所创造的智力成果依法确认并享有的权利。

由世界知识产权组织1967年7月14日签订，于1970年4月26日生效的《建立世界知识产权组织公约》对"知识产权"的范围作了如下定义：
1. 与文学、艺术及科学作品有关的权利(指版权或著作权)；
2. 与表演艺术家的表演活动、录音制品和广播节目有关的权利(指版权的邻接权)；
3. 与人类一切活动领域内的发明有关的权利(指专利权)；
4. 与发现有关的权利；
5. 与工业品外观设计有关的权利；
6. 与商品商标、服务商标、商号及其他商业标记有关的权利；
7. 与防止不正当竞争有关的权利；
8. 一切来自工业、科学及文学艺术领域的智力创作活动所产生的权利。

防火墙

防火墙是用于将因特网的子网与因特网的其余部分相隔离，以达到网络和信息安全效果的软件或硬件设施。防火墙可以被安装在一个单独的路由器中，用来过滤不想要的信息包，也可以被安装在路由器和主机中，发挥更大的网络安全保护作用。防火墙被广泛用来让用户在一个安全屏障后接入互联网，还被用来把一家企业的公共网络服务器和企业内部网络隔开。另外，防火墙还可以被用来保护企业内部网络某个部分的安全。防火墙只允许授权信息通过，而防火墙本身不能被渗透。

黑客

计算机黑客是指未经许可擅自进入某个计算机网络系统的非法用户。计算机黑客往往具有一定的计算机技术，采取截获密码等方法，非法闯入某个计算机系统，进行盗窃、修改信息、破坏系统运行等活动，对计算机网络造成很大的损失和破坏。

个人信息安全

个人信息安全是指公民身份、财产等个人信息的安全状况。随着互联网应用的普及和人们对互联网的依赖，个人信息安全受到极大的威胁。

网瘾

网瘾是指上网者由于长时间地、习惯性地沉浸在网络当中，对互联网产生强烈的依赖，以至于达到了痴迷的程度而难以自我解脱的行为状态和心理状态。上网者花费过多时间上网，以至于损害了现实中的人际关系和学业、事业。网瘾是一种心理疾病，对那些已经上网成瘾的人，应采取有针对性的教育引导、现身说法、心理辅导和心理治疗等办法，帮助他们戒除"网瘾"。

盗版软件

盗版软件是非法制造或复制的软件。它难以识别，但缺少密钥代码或组件是缺乏真实性的表现。盗版软件无保证，并且无法升级、更新。

Language Practice

I. Fill in the blanks with the missing letters of the words according to the pictures.

1. in_ _ _d

2. near-s_ _ _ted

3. c_ _ _

Unit 13 Some Important Issues

4. _ _v_ _ _nment
5. c_pyr_ _ _t
6. t_rr_b_ _
7. w_ _ _dw_de
8. f_ _ _t
9. _ddi_tion
10. tr_ff_ _ j_m

II. Complete the sentences with the words practiced above.
1. Thomas is now fighting his Internet _____.
2. Don't watch TV every night, or you will become _____.
3. Ben was startled by the _____ accident.
4. Sorry, I'm late. I was caught in the _____.
5. Lucy and her friends are searching on the net, trying to find the _____ of that disease.
6. We are the children of the earth. We should be a firm protector of our _____.
7. Everyone should abide by the _____ law.
8. I _____ to buy a new computer.
9. Premier Zhou's death caused _____ sadness.
10. Vivian is _____ for her own rights.

III. Fill in the blanks with prepositions.

Hackers (电脑黑客) **and Crackers** (解密高手)

Many people think that these two groups _____ people are the same, but actually they are not. Hackers are people, usually young people, who enter _____ other unauthorised (未授权的) system _____ fun or _____ challenge (挑战). Crackers usually do the same thing _____ hackers but _____ _____ evil will. They may try to steal technical information or to introduce some destructive (破坏性的) programs _____ other systems.

IV. Sentence completion.
1. John's overuse of the Internet _____ (使他的生活发生了巨大变化).
2. People's compulsive Internet use _____ (会影响日常生活).
3. We _____ (很费劲才完成任务) because it was so difficult.
4. Nowadays students _____ (遇到了) some important computer issues.
5. Some hackers might work out your password and _____ (闯入你的系统).

Writing

Write a short paragraph (at least 5 sentences) about some computer issues.

computer viruses ...
Internet addiction ...

Grammar

Subjunctive Mood (in if clause) (If引导的虚拟语气)	
If S + did / were ... , S+ would / might / could do ...	If I **were** you, **I would stay** there and wait for them. If I **had** a bike, I **would lend** it to you.
If S + had done ..., S + would / might / could have done ...	If I **had had** more time, I **would have checked** the paper again. If she **had been** here, she **might have helped** you.
If S + did / should / were to do ... (future time adverbial), S + would / might / could do ...	If I **saw** him tomorrow, I **would inform** him of the party. If it **should / were to** rain tomorrow, we **would cancel** our camping.
Were / Had / Should + S + done ... , ...	**Had I known** about it, I would not have taken up that much of the professor's time. **Were I to** do the work, I would do it some other way.

I. Multiple choices.

1. If the computer had not broken yesterday, he _____ his work by now.
    A. will finish B. would finish
    C. would have finished D. finish

2. _____ more careful, the document would not have been lost without saving.
    A. If the typist were B. Had the typist been

211

C. Should the typist be   D. If the typist would have been

3. If the computer _____ all the homework of the student, most of the kids would like to go to school.

    A. is to finish   B. has finished

    C. were finished   D. were to finish

4. If you _____ in good health, I _____ you go to the online friends' party yesterday.

    A. were, would have let   B. had been, would have let

    C. had been, had let   D. were, had let

5. If I knew German, I _____ his Blog to you.

    A. would have read   B. had read

    C. would read   D. will read

II. Fill in the blanks with the correct forms of the verbs in the brackets.

1. You _____ Professor Li if you went to the computer class this afternoon. (meet)

2. Had he worked harder, Louis _____ the application of spreadsheets. (grasp)

3. If the whole operation had been planned beforehand, a great deal of time and money _____. (save)

4. If you were to installed the WeChat app right now, we _____ online on Sunday. (chat)

5. We _____ the computer expert if we had been there five minutes earlier. (meet)

Word formation — Clipping (截短法)

    截短法，即将单词缩写，词义和词性保持不变，主要有截头、去尾、截头去尾等形式，如telephone截头后成为phone，laboratory 去尾后成为lab, influenza截头去尾后成为flu。

Unit 13 Some Important Issues

III. Write out the full forms of the following words.

Clipping Words	Full Form
1. quake	
2. kilo	
3. taxi	
4. exam	
5. plane	
6. maths	
7. fridge	
8. doc	

Game

Across

1. What are the important _____s of students' using computers?
2. Sometimes people feel more _____ with their online friends than the real ones.
3. Computer users should be careful about the _____ of the computers.
4. Almost all people in the world enjoy the _____s of the Internet.
5. There are some viruses that will _____ word-processing documents only.
6. Internet _____s spend a lot of time surfing the Internet, which will have a serious impact on their mental health.

Down

1. How often do others in your life _____ to you about the amount of time you spend online?
2. Internet use can also _____ with daily lifves, work and relationships with our friends.
3. I _____ to cut my online time.
4. Internet addicts seldom feel _____ about their Internet use.

 Fun Time

If Only Life Could Be Like a Computer!

If you messed up your life, you could press "Ctrl, Alt, and Delete" and start all over!

To get your daily exercise, just click on "run"! If you needed a break from life, click on "suspend".

Hit "any key" to continue life when ready.

To get even with the neighbors, turn up the sound blaster.

To add / remove someone in your life, click settings and control panel.

To improve your appearance, just adjust the display settings.

If life gets too noisy, turn off the speakers.

When you loose your car keys, click on "find".

"Help" with the chores is just a click away.

Auto insurance wouldn't be necessary. You would use your diskette to recover from a crash.

And, we could click on "SEND NOW" and a Pizza would be on its way to you …

Unit 13 Some Important Issues

Project

Communicate with others!

Step 1: Divide the class into 6 groups. Ask 3 groups to talk about the benefits of the Internet we enjoy.

Step 2: Ask another 3 groups to talk about the problems of the Internet.

Step 3: Choose 3 students respectively from groups with different subjects to have a debate on the advantages and disadvantages of the Internet.

Self-checklist

根据实际情况，从A、B、C、D中选择合适的答案：A代表你能很好地完成该任务；B代表你基本上可以完成该任务；C代表你完成该任务有困难；D代表你不能完成该任务。

A B C D

☐ ☐ ☐ ☐ 1. 能掌握并能运用本单元所学重点句型、词汇和短语。

☐ ☐ ☐ ☐ 2. 能理解并正确模仿听说部分的句子，正确掌握发音及语调。

☐ ☐ ☐ ☐ 3. 能模仿句型进行简单的对话。

☐ ☐ ☐ ☐ 4. 能读懂本课的短文，并正确回答相关问题。

☐ ☐ ☐ ☐ 5. 能掌握由if引导的虚拟语气的用法。

☐ ☐ ☐ ☐ 6. 能掌握由if引导的非真实条件句的三种情况。

☐ ☐ ☐ ☐ 7. 能掌握截短词的构词方法。

☐ ☐ ☐ ☐ 8. 能用课文中学习的词组造句。

☐ ☐ ☐ ☐ 9. 能向同学介绍使用电脑所带来的常见问题。

Unit 14 Future Computers

Unit Goals

In this unit, you will be able to
- understand future computers;
- talk about computers in the future;
- understand articles about the future of computers;
- write a short paragraph about future computers;
- demonstrate your vision about future computers.

Lead-in

Picture matching.

1. tech support
2. computer trainer
3. webmaster
4. programmer
5. network manager
6. industry designer

Unit 14 Future Computers

Listening and Speaking

Listen and complete.

a. Hello, what do you think the future computers will be like?
b. I really hope that computers in the future can read people's minds.
c. I wish future computers would have virtual screens.
d. They will save energy and are always ready for use.
e. One that can do the housework, you know, do the cleaning and cooking, and look after the kids ...

Computers in the Future

A magazine reporter from *Computer World* is going to write an article about future computers. Now he is interviewing people and making a survey.

He is talking to a famous writer, Mr Rowley.

❶ Reporter: Excuse me, sir. What do you think the future computers will be like?
Mr Rowley: I'm an author. _____
Reporter: Why do you want to have a computer like that?
Mr Rowley: Because it can help record my ideas.

He turns to Mrs Rowley, a housewife.

❷ Reporter: What kind of computers do you want to have in the future,

ma'am?
 Mrs Rowley: _____
  Reporter: What will you do, then?
 Mrs Rowley: Keep the remote-control unit in hand, of course.

He is talking to Billy, a student.
❸ Reporter: _____
   Billy: I wish computers would be solar-powered. Whenever there is light, there is electricity supply.
  Reporter: Sounds fantastic. They are energy-saving.
   Billy: Yeah. _____

He is talking to Charles, a boy from a kindergarten.
❹ Reporter: What is the dream computer for you, kid?
   Charles: _____
  Reporter: And then?
   Charles: They will allow me to play computer games in virtual reality.

Role Play

A: What do you think the future computers will be like?
B: Well, they will be really smart. They can ...

Words & Expressions

author /ˈɔːθə(r)/ *n.* 作家，作者
remote-control /rɪˈməʊt kənˈtrəʊl/ *n.* 遥控；遥控器

kindergarten /ˈkɪndəɡɑːtn/ *n.* 幼儿园

look after 照顾

Unit 14 Future Computers

Text A

Pre-reading activities.

1. What will computers be like in five years' time?

2. How will they affect our life?

Reading Strategy

识别文章的不同体裁 Identifying Different Genres of Writing 不同体裁的文章，如记叙文（Narration）、说明文（Exposition）、议论文（Argumentation）、应用文（Practical Writing）等，具有不同的特征。作者的写作手法不同，表达的方式千差万别。体裁不同，要求使用的阅读方法也不一样。因此，掌握不同体裁文章的阅读方法有助于提升我们的阅读速度、提高我们的阅读效率、加深我们对文章的理解。

Future Computers

No one knows what computers will look like and what they can do in the future. But one thing is certain, as long as people think of ways to use computers they will follow the orders. In other words, imagination makes inventions.

Household computer systems may well make our lives different. In the future people will be able to set room temperature, make food ready, draw curtains and record TV programs from anywhere in the world using their mobile phones.

In the future people do not have to travel to work because they can work from home. It means offices will be smaller, and e-commerce will be more popular.

In the future the technology of artificial intelligence will be further developed and may greatly change the lifestyles of people. Computers will also look after themselves more, carrying out maintenance tasks automatically.

Words & Expressions

imagination /ɪˌmædʒɪˈneɪʃn/ *n.* 想象
invention /ɪnˈvenʃn/ *n.* 发明，创造
temperature /ˈtemprətʃə(r)/ *n.* 温度
curtain /ˈkɜːtn/ *n.* 窗帘，门帘
artificial /ˌɑːtɪˈfɪʃl/ *adj.* 人工的；人造的

intelligence /ɪnˈtelɪdʒəns/ *n.* 智能；智力
further /ˈfɜːðə(r)/ *adv.* 进一步
maintenance /ˈmeɪntənəns/ *n.* 维护；保持

in other words 换句话说

True or false.
1. In the future, computers will do things the way we want them to do.
2. Computers can help us do all our housework.
3. Businesses will be done electronically.
4. Offices will be of no use in the future because people will work from home.
5. Mobile phones will be more powerful in the future.

Text B

Pre-reading questions.
1. Have you ever heard of cloud computing?
2. Can you list some applications of cloud computing?

Cloud Computing

Everyone is talking about "the cloud". But what does it mean?

Cloud computing is Internet-based computing, through which shared resources, software and information are used by computers and other devices on demand. Cloud computing describes a new model for IT service.

From the business perspective, there are numerous reasons to use cloud computing, some of the most common being payment for only what you use (and not wasting resources) and easy deployments to end-users.

You're probably already in the cloud right now — Baidu Cloud, AliCloud — since SaaS (software as a service) applications are all supported by cloud computing.

Cloud computing is a revolutionary concept. It begins to change the way people work and live.

Words & Expressions

demand /dɪ'mɑːnd/ *n.* 要求
model /'mɒdl/ *n.* 模式，模型
perspective /pə'spektɪv/ *n.* 看法；视角
deployment /dɪ'plɔɪmənt/ *n.* 展开；部署
revolutionary /ˌrevə'luːʃənəri/ *adj.* 革命的

concept /'kɒnsept/ *n.* 概念，观念

on demand 一经请求
right now 立即

Short-answer questions.

Answer the following questions according to Text B.

1. What does the word "cloud" refer to?

2. What are used by computers and other devices through cloud computing?

   _____

3. What is cloud computing based on?

   _____

4. How do people think of the concept of cloud computing?

   _____

5. What will cloud computing change?

   _____

Notes

artificial intelligence 人工智能：人工智能，英文缩写为AI。它是研究、开发用于模拟、延伸和扩展人的智能的理论、方法、技术及应用系统的一门新的技术科学。人工智能是计算机科学的一个分支，它企图了解智能的实质，并生产出一种新的能以人类智能相似的方式做出反应的智能机器，该领域的研究包括机器人、语言识别、图像识别、自然语言处理和专家系统等。人工智能不是人的智能，但能像人那样思考、也可能超过人的智能。

smart home 智能家居：又称智能住宅。通俗地说，它是利用先进的计算机、嵌入式系统和网络通信技术，将家中的各种设备通过家庭网络连接到一起。一方面，智能家居将让用户有更方便的手段来管理家庭设备；另一方面，智能家居内的各种设备相互间可以通讯，不需要用户指挥也能根据不同的状态互动运行，从而给用户带来最大程度的高效、便利、舒适与安全。

cloud computing 云计算：是分布式计算技术的一种，其最基本的概念，是通过网络将庞大的计算处理程序自动拆分成无数个较小的子程序，再交由多部服务器所组成的庞大系统，经搜寻、计算分析之后将处理结果回传给用户。

通过这项技术，网络服务提供者可以在数秒之内处理数以千万计甚至亿计的信息，达成和"超级计算机"同样强大效能的网络服务。"

SaaS 软件运营：简称为软营，将应用软件统一部署在云服务器上为用户提供软件服务。软营公司为企业搭建信息化所需要的所有网络基础设施及软件、硬件运作平台，并负责所有前期的实施、后期的运营维护等一系列服务，企业无需购买软硬件、建设机房、招聘IT人员，即可通过互联网使用信息系统。

语音识别

　　语音识别就是研究让机器最终能听懂人类口述的自然语言的一门学科。听懂有两种含义：第一种是将这种口述语言逐词(字)逐句地转换为相应的文字，例如对口述文章作听写；第二种则是对口述语言中所包含的要求或询问作出正确的反应，而不拘泥于把所有词正确转换为书面文字。语音识别和语音合成相结合，即构成一个完整的"人机对话通信系统"。

借记卡和信用卡

　　借记卡是让持卡人花自己事先存入的钱，即"先存钱，后消费"；而信用卡则是花银行的钱，即"先消费，后还款"。可以说，信用卡是个人申请信用贷款的一种方式，是帮助持卡人实现资金短期周转、取代现金的支付工具。

支付宝(AliPay)

　　支付宝(中国)网络技术有限公司是国内第三方支付平台，致力于提供"简单、安全、快速"的支付解决方案。支付宝公司旗下有"支付宝"与"支付宝钱包"两个独立品牌。支付宝主要提供支付及理财服务，包括网购担保交易、网络支付、转账、信用卡还款、手机充

223

值、水电煤缴费、个人理财等多个领域。在进入移动支付领域后，为零售百货、电影院线、连锁商超和出租车等多个行业提供服务，还推出了余额宝等理财服务。支付宝与国内外180多家银行以及VISA、MasterCard国际组织等机构建立了合作关系。

网际协议安全 (IP Security)

IPSec作为安全网络的长期方向，是基于密码学的保护服务和安全协议的套件。

大数据 (Big Data)

指无法在一定时间范围内用常规软件工具进行捕捉、管理和处理的数据集合，是需要新处理模式才能具有更强的决策力、洞察发现力和流程优化能力的海量、高增长率和多样化的信息资产。

物联网 (IOT)

物联网是新一代信息技术的重要组成部分。物联网的英文名称叫"The Internet of Things"。顾名思义，物联网就是"物物相连的互联网"，是将各种信息传感设备与互联网结合起来而形成的一个巨大网络。

Language Practice

I. Fill in the blanks with the missing letters of the words according to the pictures.

1. tr_ _ner

2. m_d_l

3. h_ _sew_ _k

Unit 14 Future Computers

4. r_p_ _ter

5. n_mer_ _s

6. c_ _t_ _ns

7. _ _tific_ _l _ _tellig_ _ _ _

8. TV pr_gr_ _

9. r_m_te-c_ _trol

10. t_ _per_ture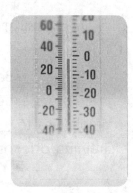

II. Complete the sentences with the words practiced above.

1. The most exciting _____ tonight is the football match.
2. As a _____, you must be alert to what takes place around you.
3. There are _____ sorts of software to choose from.
4. A _____ is one who trains, while a trainee is one who is trained.
5. With this _____ unit, you can lock your car, even if you are 5 meters away from the car.
6. Please open the _____. I want to be bathed in sunlight.
7. Robots are typical examples of _____.

225

8. The _____ of today is between 2℃ ~ 8℃.
9. This is the _____ of the new building.
10. Some housewives complain that it is too boring to do _____ every day.

Ⅲ. Fill in the blanks with the words in the box.

| those   support   industry   business   life   date |

Training and Jobs

People in the IT _____ change jobs frequently, and IT skills become out of _____ very quickly. Few people now expect to have jobs for _____. Because of these, companies often have to train newcomers. Training is expensive. Companies that do not give training often lose employees to _____ that do. Training is itself a big _____. Training often costs more than the computer system that is designed to _____.

Ⅳ. Sentence completion.
1. The research we did was _____ (以你的数据库中的数据为根据的).
2. Julia has three children to _____ (照顾).
3. This antivirus software is _____ (目前很受欢迎).
4. It is going to rain tomorrow. _____ (也就是说), I am not going to attend the lecture on cloud computing.
5. These books are printed _____ (应要求为计算机爱好者).

226

Unit 14 Future Computers

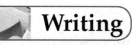

Writing

Write a short paragraph (at least 5 sentences) about your imagination of future computer world.

Hints:
household computer systems: set room temperature, make food ready, draw curtains, record TV programs, using mobile phones.
work from home
e-commerce
carry out maintenance tasks automatically

Grammar

Subjunctive Mood (in other cases) (其它情况下的虚拟语气)	
suggest, advise, insist, urge ...	They suggested that everyone **(should) have** a copy.
It is important / necessary ... that ...	It is necessary that we **(should) arrive** in time.
wish	I wish I **knew** the truth. I wish I **had taken** your advice.
If only	If only he **were** here. If only he **hadn't driven** so fast.
as if / though	John felt as if he alone **were** responsible for the accident. He talks as if he **had been** there at the time.
It is / was (about / high) time that ...	It is high time that we **left**.

227

I. Multiple choices.

1. Bob wishes that he _____ computer science instead of history when he was in the university.
   A. studies   B. studied
   C. has studied   D. had studied

2. The teacher suggests that the students _____ two weeks to finish their program.
   A. spend   B. would spend
   C. be spent   D. would be spent

3. Martin is very good at software, as if he _____ a student in computer science department.
   A. were   B. would be
   C. have been   D. would have been

4. It is important that Linda _____ the Internet for an hour every day.
   A. surf   B. surfs
   C. must have   D. would surf

5. It is time that I _____ with my boyfriend online.
   A. chat   B. chatted
   C. have chatted   D. will chat

II. Fill in the blanks with the correct forms of the verbs in the brackets.

1. It is about time you _____ the computer class. (attend)
2. If only I _____ your e-mail before I started my work! (read)
3. Li Ming acts the way as though he _____ a computer expert. (be)
4. It is essential that Tom _____ a PowerPoint file for his lecture. (prepare)
5. Bob ordered that the document _____ at once. (save)

Word formation — Acronymy (词首字母缩略法)

词首字母缩略法，即用单词首字母组成一个新词。读音主要有两种形式，即各字母分别读音，如：C.O.D., cash on delivery (货到即付)；或作为一个单词读音，如：SALT, Strategic Arms Limitation Talks (限制战略武器会谈)。

Unit 14 Future Computers

III. Write out the full forms of the following words.

A	B
1. VIP	
2. DIY	
3. WWW	
4. GPS	
5. 5G	
6. AI	

Game

Across

1. This laptop computer is the latest _____.
2. Cloud computing is a new _____ to most people.
3. The _____s in this room can come down automatically because they are controlled by computers.
4. There are _____ women who did very well in IT field.
5. Experienced _____s may design your clothes according to your demands.

Down

1. In the future the technology of _____ intelligence may greatly change the lifestyles of people.

229

2. The _____ of the computer brought great changes to human life.
3. If you want to be a _____, you have to be good at web page design and management.
4. Ice cream should be kept at very low _____.
5. The electronic store imported some latest models of computers to meet consumers' _____s.

 Fun Time

I Have Windows

A blonde walked into a store to buy curtains.

She went up to the salesman and said, "I want those pink curtains to fit my computer screen."

The salesman mentioned, "Computers don't need curtains."

The blonde said, "Hello? I have windows!"

 Project

Communicate with others!

Step 1: Interview some classmates to learn what computers will be like in the future in their mind.
Step 2: Write an article about future computers based on the interview.
Step 3: Ask three classmates to read their writings in class and invite others to comment on.

Self-checklist

根据实际情况，从A、B、C、D中选择合适的答案：A代表你能很好地完成该任务；B代表你基本上可以完成该任务；C代表你完成该任务有困难；D代表你不能完成该任务。

A	B	C	D	
☐	☐	☐	☐	1. 能掌握并能运用本单元所学重点句型、词汇和短语。
☐	☐	☐	☐	2. 能理解并正确模仿听说部分的句子，正确掌握发音及语调。
☐	☐	☐	☐	3. 能模仿句型进行简单的对话。
☐	☐	☐	☐	4. 能读懂本课的短文，并正确回答相关问题。
☐	☐	☐	☐	5. 能掌握其他虚拟语气用法。
☐	☐	☐	☐	6. 能掌握虚拟语气的特殊变化。
☐	☐	☐	☐	7. 能掌握词首字母缩略词的构词方法。
☐	☐	☐	☐	8. 能用课文中学习的词组造句。
☐	☐	☐	☐	9. 能向同学介绍对未来电脑世界的设想。

Words & Expressions

A

a piece of cake		很容易，小菜一碟	(Unit 9)
a variety of		多种的，各种各样的	(Unit 11)
academic /ˌækəˈdemɪk/	adj.	学术上的	(Unit 3)
access /ˈækses/	n.	通路；入门	
	v.	存取；接近	(Unit 2)
according to		按照，根据	(Unit 11)
account /əˈkaʊnt/	n.	账目	(Unit 6)
activate /ˈæktɪveɪt/	v.	激活	(Unit 8)
activity /ækˈtɪvəti/	n.	活动	(Unit 12)
actual /ˈæktʃuəl/	adj.	实际的	(Unit 12)
add up		加起来	(Unit 6)
addict /ˈædɪkt/	n.	对……入迷、上瘾的人	(Unit 13)
addiction /əˈdɪkʃn/	n.	沉溺，上瘾	(Unit 13)
addictive /əˈdɪktɪv/	adj.	易上瘾的，使人入迷的	(Unit 10)
advance /ədˈvɑːns/	n. & v.	前进；进步，发展	(Unit 3)
advancement /ədˈvɑːnsmənt/	n.	进步	(Unit 8)
advertise /ˈædvətaɪz/	v.	做广告，登广告	(Unit 12)
afford /əˈfɔːd/	v.	买得起；承担	(Unit 1)
all of a sudden		忽然，一下子	(Unit 6)
amazing /əˈmeɪzɪŋ/	adj.	令人惊异的	(Unit 7)
analysis /əˈnæləsɪs/	n.	分析	(Unit 6)
angle /ˈæŋgl/	n.	角，角度	(Unit 2)
annual /ˈænjuəl/	adj.	每年的	(Unit 8)

antivirus /ˈæntivaɪrəs/	adj.	抗病毒	(Unit 10)
appliance /əˈplaɪəns/	n.	器具，器械	(Unit 10)
application /ˌæplɪˈkeɪʃn/	n.	应用程序；应用；申请	(Unit 3)
apply to		应用于	(Unit 9)
arrange /əˈreɪndʒ/	v.	安排	(Unit 8)
article /ˈɑːtɪkl/	n.	文章；论文	(Unit 5)
artificial /ˌɑːtɪˈfɪʃl/	adj.	人工的；人造的	(Unit 14)
as ... as		和……一样，像……一样	(Unit 2)
as is known		众所周知	(Unit 9)
assignment /əˈsaɪnmənt/	n.	作业	(Unit 5)
assistant /əˈsɪstənt/	n.	助手；助教	(Unit 3)
attach /əˈtætʃ/	v.	附上，贴上	(Unit 2)
attachment /əˈtætʃmənt/	n.	附件	(Unit 12)
attribute /ˈætrɪbjuːt/	n.	属性，特性	(Unit 9)
audience /ˈɔːdiəns/	n.	观众，听众	(Unit 8)
author /ˈɔːθə(r)/	n.	作家，作者	(Unit 14)
authorised /ˈɔːθəraɪzd/	adj.	权威认可的，经授权的	(Unit 11)
automatically /ˌɔːtəˈmætɪkli/	adv.	自动地	(Unit 6)
available /əˈveɪləbl/	adj.	可用的	(Unit 7)
avoid /əˈvɔɪd/	v.	避免	(Unit 9)
avoid doing		避免做某事	(Unit 9)

B

background /ˈbækɡraʊnd/	n.	背景，后台	(Unit 3)
bar code		条形码	(Unit 10)
base /beɪs/	v.	以……为基础	(Unit 11)
base on / upon		以……为基础	(Unit 11)
basic /ˈbeɪsɪk/	n.	(通常用复数) 基本因素，基本原理	(Unit 4)
be afraid (that)		害怕，担心	(Unit 9)
be busy doing sth		忙于做某事	(Unit 8)
be different from		与……不同	(Unit 5)

Words & Expressions

be easy to use		易于使用	(Unit 8)
be in touch with		与……联系	(Unit 12)
be similar to		与……相似	(Unit 12)
because of		因为	(Unit 11)
benefit /'benɪfɪt/	n.	益处，好处	(Unit 13)
block /blɒk/	n.	（一）批	(Unit 5)
Board of Directors		董事会	(Unit 5)
bold /bəʊld/	adj.	粗体的	
	n.	黑体字，粗体字	(Unit 5)
branch /brɑːntʃ/	n.	枝，分枝；分部	(Unit 11)
branded /'brændɪd/	adj.	属于品牌的	(Unit 2)
break down		毁掉；停顿；中止	(Unit 11)
break into		闯入	(Unit 13)
bring ... in		吸引；带入	(Unit 8)
bring about		使发生，致使	(Unit 13)
brochure /'brəʊʃə(r)/	n.	小册子	(Unit 9)
browser /'braʊzə(r)/	n.	浏览器	(Unit 3)
budget /'bʌdʒɪt/	n.	预算	(Unit 4)
built-in /ˌbɪlt'ɪn/	adj.	内置的	(Unit 1)
burst /bɜːst/	v.	爆发	(Unit 6)
burst into laughter		突然大笑	(Unit 11)
bus /bʌs/	n.	（计算机系统的）总线	(Unit 2)
button /'bʌtn/	n.	按钮；钮扣	(Unit 4)

C

cable /'keɪbl/	n.	电缆	(Unit 1)
calculate /'kælkjuleɪt/	v.	计算	(Unit 6)
calculation /ˌkælkjʊ'leɪʃn/	n.	计算	(Unit 4)
calorie /'kæləri/	n.	卡路里 (热量单位)	(Unit 6)
carry around		四处携带	(Unit 1)
catalogue /'kætəlɒg/	n.	目录；目录册	(Unit 12)

235

cell /sel/	n.	单元格	(Unit 6)
central /ˈsentrəl/	adj.	中央的	(Unit 1)
channel /ˈtʃænl/	n.	通道；渠道，途径；频道	(Unit 11)
character /ˈkærəktə(r)/	n.	文字	(Unit 7)
chart /tʃɑːt/	n.	图表	(Unit 8)
check in		签到；(旅馆、机场等)登记	(Unit 7)
check out		付账离开，结账	(Unit 7)
chime /tʃaɪm/	v.	插嘴	(Unit 10)
chime in		插话	(Unit 10)
circuit /ˈsɜːkɪt/	n.	电路	(Unit 2)
civilisation /ˌsɪvəlaɪˈzeɪʃn/	n.	文明	(Unit 10)
classify /ˈklæsɪfaɪ/	v.	分类	(Unit 11)
clerk /klɑːk/	n.	职员	(Unit 6)
click /klɪk/	v.	点击	(Unit 4)
clinic /ˈklɪnɪk/	n.	诊所	(Unit 7)
collect /kəˈlekt/	v.	收集，搜集	(Unit 7)
collection /kəˈlekʃn/	n.	一批物品	(Unit 11)
column /ˈkɒləm/	n.	圆柱；列	(Unit 6)
combine /kəmˈbaɪn/	v.	使结合	(Unit 9)
come across		遇到	(Unit 13)
comfortable /ˈkʌmftəbl/	adj.	舒适的	(Unit 13)
commercial /kəˈmɜːʃl/	adj.	商业的	(Unit 7)
commit /kəˈmɪt/	v.	犯(错误)，做(坏事)	(Unit 10)
communicate /kəˈmjuːnɪkeɪt/	v.	交流	(Unit 8)
complain /kəmˈpleɪn/	v.	抱怨	(Unit 13)
complex /ˈkɒmpleks/	adj.	复杂的	(Unit 9)
component /kəmˈpəʊnənt/	n.	元件，部件；成分	(Unit 2)
compulsive /kəmˈpʌlsɪv/	adj.	上瘾的，无法控制行为的	(Unit 13)
concept /ˈkɒnsept/	n.	概念，观念	(Unit 14)
confused /kənˈfjuːzd/	adj.	混淆的；困惑的	(Unit 12)
connect /kəˈnekt/	v.	连接	(Unit 1)

Words & Expressions

consequence	/'kɒnsɪkwəns/	n.	结果	(Unit 13)
consist	/kən'sɪst/	v.	组成	(Unit 8)
consist of			由……组成	(Unit 8)
consistent	/kən'sɪstənt/	adj.	一致的	(Unit 9)
construction	/kən'strʌkʃn/	n.	建设	(Unit 11)
contain	/kən'teɪn/	v.	包含	(Unit 8)
convenient	/kən'viːniənt/	adj.	方便的；省事的	(Unit 1)
countless	/'kaʊntləs/	adj.	无数的	(Unit 5)
CPU (Central Processing Unit)		n.	中央处理器	(Unit 1)
credit	/'kredɪt/	n.	信用	(Unit 10)
crime	/kraɪm/	n.	罪恶	(Unit 10)
curtain	/'kɜːtn/	n.	窗帘，门帘	(Unit 14)

D

damage	/'dæmɪdʒ/	n.	损害，毁坏	(Unit 13)
data	/'deɪtə/	n.	(datum 的复数) 资料，数据	(Unit 2)
database	/'deɪtəbeɪs/	n.	数据库，资料库	(Unit 3)
date	/deɪt/	n.	日期	(Unit 7)
decade	/'dekeɪd/	n.	十年	(Unit 8)
definitely	/'defɪnətli/	adv.	明确地，干脆地	(Unit 7)
delete	/dɪ'liːt/	v.	删除	(Unit 5)
deliver	/dɪ'lɪvə(r)/	v.	陈述，发言；递送	(Unit 10)
demand	/dɪ'mɑːnd/	n.	要求	(Unit 14)
dental	/'dentl/	adj.	牙齿的	(Unit 7)
deployment	/dɪ'plɔɪmənt/	n.	展开；部署	(Unit 14)
designated	/'dezɪgneɪtɪd/	adj.	指定的	(Unit 8)
designer	/dɪ'zaɪnə(r)/	n.	设计师	(Unit 3)
desktop computer			台式机	(Unit 1)
device	/dɪ'vaɪs/	n.	装置，设备	(Unit 2)
devise	/dɪ'vaɪz/	v.	设计	(Unit 7)
digital pen			数字笔	(Unit 1)

direction /dɪ'rekʃn, daɪ'rekʃn/	n.	方向	(Unit 11)
disadvantage /ˌdɪsəd'vɑːntɪdʒ/	n.	劣势，短处	(Unit 10)
disappear /ˌdɪsə'pɪə(r)/	v.	消失	(Unit 13)
display /dɪ'spleɪ/	v. & n.	陈列；展览；显示	(Unit 4)
distance /'dɪstəns/	n.	距离，远程	(Unit 10)
divide /dɪ'vaɪd/	v.	[数] 除	(Unit 6)
do a survey		做调查	(Unit 8)
do sb a favour		帮某人忙	(Unit 9)
document /'dɒkjʊmənt/	n.	文件	(Unit 5)
double-click	n. & v.	双击	(Unit 4)
doubt /daʊt/	v.	怀疑	(Unit 12)
download /ˌdaʊn'ləʊd/	v.	下载	(Unit 3)
dynamic /daɪ'næmɪk/	adj.	动态的	(Unit 8)
dynasty /'dɪnəsti/	n.	朝代	(Unit 8)

E

e-bank /'iːbæŋk/	n.	电子银行	(Unit 12)
e-commerce /iː'kɒmɜːs/	n.	电子商务	(Unit 12)
e-pal /'iːpæl/	n.	网友	(Unit 12)
edit /'edɪt/	v.	编辑，校订	
	n.	编辑工作	(Unit 5)
education /ˌedʒu'keɪʃn/	n.	教育	(Unit 10)
electronic /ɪˌlek'trɒnɪk/	adj.	电子的	(Unit 12)
element /'elɪmənt/	n.	成分；要素	(Unit 9)
enable sb to do sth		使某人能做某事	(Unit 10)
end-user /ˌend'juːzə(r)/	n.	终端用户	(Unit 3)
error /'erə(r)/	n.	错误	(Unit 6)
Ethernet /'iːθənet/		以太网，采用共享总线型传输媒体方式的局域网	(Unit 11)
evil /'iːvl/	adj.	邪恶的	(Unit 13)
excel /ɪk'sel/	v.	优于，擅长	(Unit 9)

Words & Expressions

excel at		(在某方面) 出色	(Unit 9)
exit /'eksɪt/	n.	退出；出口	
	v.	离开当前命令；退出	(Unit 4)
expert /'ekspɜːt/	n.	专家	(Unit 4)
expire /ɪk'spaɪə(r)/	v.	期满，终止	(Unit 3)
export /'ekspɔːt/	n.	输出，出口	(Unit 7)
extensively /ɪk'stensɪvli/	adv.	广泛地	(Unit 11)
eye-catching /'aɪ kætʃɪŋ/	adj.	引人注目的	(Unit 9)
facilitate /fə'sɪlɪteɪt/	v.	使便利；促进；为他人提供方便 (或机会)	(Unit 11)

F

fibre /'faɪbə(r)/	n.	光纤；纤维	(Unit 11)
fight /faɪt/	v.	战斗；作战	(Unit 13)
figure out		想出；计算出	(Unit 1)
finance /faɪ'næns/	n.	财政，金融	(Unit 3)
financial /faɪ'nænʃl/	a.	财政的，金融的	(Unit 4)
firm /fɜːm/	n.	公司	(Unit 9)
fix /fɪks/	v.	修理，修复	(Unit 13)
folder /'fəʊldə(r)/	n.	文件夹	(Unit 4)
font /fɒnt/	n.	字体，字形	(Unit 5)
format /'fɔːmæt/	v.	格式化 (磁盘)	(Unit 4)
format /'fɔːmæt/	n.	格式	(Unit 8)
format /'fɔːmæt/	v.	排版，版式设计	
	n.	版面	(Unit 9)
formula /'fɔːmjələ/	n.	(pl. formulas 或 formulae) 公式；规则	(Unit 6)
frame /freɪm/	n.	框架，文本框	(Unit 9)
free of charge		免费	(Unit 3)
freeware /'friːweə(r)/	n.	免费软件	(Unit 3)
function /'fʌŋkʃn/	n.	函数	(Unit 6)
further /'fɜːðə(r)/	adv.	进一步	(Unit 14)

furthermore /ˌfɜːðəˈmɔː(r)/	adv.	而且，此外	(Unit 10)

G

generally /ˈdʒenrəli/	adv.	通常，大体上	(Unit 9)
generally speaking		一般说来，通常	(Unit 11)
get rid of		除掉	(Unit 5)
Good luck!		祝你好运！	(Unit 9)
graph /grɑːf/	n.	图表；曲线图	(Unit 6)
graphic /ˈɡræfɪk/	adj.	绘画的，图解的	(Unit 3)
graphics /ˈɡræfɪks/	n.	图案	(Unit 8)
grocery /ˈɡrəʊsəri/	n.	<美>食品杂货店；食品，杂货	(Unit 7)
GUI		图形用户界面 (Graphical User Interface，又称图形用户接口)	(Unit 4)
guilty /ˈɡɪlti/	adj.	内疚的，有罪的	(Unit 13)

H

hack /hæk/	v.	劈，砍；破坏	(Unit 10)
hacker /ˈhækə(r)/	n.	黑客	(Unit 10)
hand in		上交	(Unit 11)
handwriting recognition		手写识别	(Unit 1)
have trouble in		在……有困难	(Unit 13)
hear of		听说	(Unit 6)
highlight /ˈhaɪlaɪt/	v.	选中；使显著	(Unit 6)
household /ˈhaʊshəʊld/	adj.	家用的，家庭的	(Unit 10)
humour /ˈhjuːmə(r)/	n.	幽默	(Unit 6)

I

icon /ˈaɪkɒn/	n.	图标；肖像	(Unit 4)
identification (ID) /aɪˌdentɪfɪˈkeɪʃn/	n.	身份	(Unit 6)
idiot /ˈɪdiət/	n.	傻瓜	(Unit 6)
image /ˈɪmɪdʒ/	n.	图像	(Unit 8)

Words & Expressions

imagination /ɪˌmædʒɪˈneɪʃn/	n.	想象	(Unit 14)
immediately /ɪˈmiːdiətli/	adv.	立即，马上	(Unit 6)
import /ˈɪmpɔːt/	n.	输入，进口	(Unit 7)
in case		以免，万一	(Unit 5)
in charge of		负责，主管	(Unit 3)
in fact		事实上	(Unit 1)
in front of		在……前面	(Unit 10)
in order to		为了	(Unit 5)
in other words		换句话说	(Unit 14)
include /ɪnˈkluːd/	v.	包括	(Unit 9)
indispensable /ˌɪndɪˈspensəbl/	adj.	不可缺少的	(Unit 5)
individual /ˌɪndɪˈvɪdʒuəl/	adj.	单独的；个人的	(Unit 8)
infect /ɪnˈfekt/	v.	感染；传染	(Unit 13)
input /ˈɪnpʊt/	n. & v.	输入	(Unit 2)
insert /ɪnˈsɜːt/	v.	插入	(Unit 5)
install /ɪnˈstɔːl/	v.	安装	(Unit 3)
instance /ˈɪnstəns/	n.	实例	(Unit 8)
integrate /ˈɪntɪɡreɪt/	v.	使结合成为整体	(Unit 8)
intellectual /ˌɪntəˈlektʃuəl/	adj.	知识的；智力的	
	n.	知识分子	(Unit 13)
intellectual property		知识产权	(Unit 13)
intelligence /ɪnˈtelɪdʒəns/	n.	智能；智力	(Unit 14)
intend /ɪnˈtend/	v.	打算	(Unit 13)
interface /ˈɪntəfeɪs/	n.	界面	(Unit 12)
interfere /ˌɪntəˈfɪə(r)/	v.	干预，妨碍	(Unit 13)
interfere with		打扰；妨碍	(Unit 13)
interview /ˈɪntəvjuː/	v.	采访	(Unit 6)
introduce /ˌɪntrəˈdjuːs/	v.	介绍；传入，引进	(Unit 2)
introduction /ˌɪntrəˈdʌkʃn/	n.	介绍	(Unit 4)
invention /ɪnˈvenʃn/	n.	发明，创造	(Unit 14)
inventory /ˈɪnvəntri/	n.	详细目录；库存	(Unit 7)

241

involve /ɪnˈvɒlv/	v.	包含	(Unit 4)
isolate /ˈaɪsəleɪt/	v.	使孤立，隔离	(Unit 13)
issue /ˈɪʃuː/	n.	问题	(Unit 13)
It's a great help to me.		它对我的帮助太大了。	(Unit 1)
It's really fun.		它真的很有趣。	(Unit 1)
italic /ɪˈtælɪk/	adj.	斜体的	
	n.	斜体	(Unit 5)

J

journalist /ˈdʒɜːnəlɪst/	n.	新闻记者	(Unit 5)

K

keep a record of		记录	(Unit 6)
keep track of		跟上……的	(Unit 2)
keep track of		了解……的动态	(Unit 6)
keyboard /ˈkiːbɔːd/	n.	键盘	(Unit 1)
kindergarten /ˈkɪndəɡɑːtn/	n.	幼儿园	(Unit 14)
known as		被称为；以……著称	(Unit 2)

L

label /ˈleɪbl/	n.	标签；商标	(Unit 6)
LAN /læn/		网络，局域网，本地网	(Unit 11)
laptop computer		笔记本电脑	(Unit 1)
laser /ˈleɪzə(r)/	n.	激光	(Unit 2)
laughter /ˈlɑːftə(r)/	n.	笑；笑声	(Unit 6)
lay /leɪ/	v.	放置，平铺	(Unit 9)
layout /ˈleɪaʊt/	n.	布局，安排	(Unit 8)
lecture /ˈlektʃə(r)/	n.	演讲，讲课	(Unit 10)
librarian /laɪˈbreəriən/	n.	图书管理员	(Unit 10)
license /ˈlaɪsns/	n.	许可证，执照	(Unit 3)
license holder		许可证持有者	(Unit 3)

Words & Expressions

limitless /ˈlɪmɪtləs/	adj.	无限的	(Unit 6)
link /lɪŋk/	v.	联合，联合	(Unit 7)
locate /ləʊˈkeɪt/	v.	定位	(Unit 5)
logo /ˈləʊgəʊ/	n.	标态，商标	(Unit 9)
look after		照顾	(Unit 14)
lose contact with		和……失去联系	(Unit 10)
luckily /ˈlʌkɪli/	adv.	幸运地	(Unit 13)

M

mail /meɪl/	n.	邮件	(Unit 12)
main /meɪn/	adj.	主要的	(Unit 12)
maintenance /ˈmeɪntənəns/	n.	维护；保持	(Unit 14)
make a difference		使不同	(Unit 8)
make up of		由……构成，组成	(Unit 12)
male /meɪl/	n.	男性	
	adj.	男的；雄性的	(Unit 7)
manual /ˈmænjʊəl/	adj.	手动的，手工的	(Unit 7)
margin /ˈmɑːdʒɪn/	n.	页边的空白	(Unit 5)
marketability /ˌmɑːkɪtəˈbɪləti/	n.	可销售性，市场性	(Unit 3)
mass produce		批量生产	(Unit 1)
master /ˈmɑːstə(r)/	n.	母版	(Unit 9)
mechanically /məˈkænɪkli/	adv.	机械地	(Unit 8)
mention /ˈmenʃn/	v.	提及	(Unit 4)
messenger /ˈmesɪndʒə(r)/	n.	报信者；邮递员	(Unit 12)
microprocessor /ˌmaɪkrəʊˈprəʊsesə(r)/	n.	微处理器	(Unit 2)
model /ˈmɒdl/	n.	模式，模型	(Unit 14)
modify /ˈmɒdɪfaɪ/	v.	修改	(Unit 4)
monitor /ˈmɒnɪtə(r)/	n.	显示器	(Unit 1)
moreover /mɔːrˈəʊvə(r)/	adv.	而且	(Unit 10)
motherboard /ˈmʌðəbɔːd/	n.	主板，母板	(Unit 2)
mouse /maʊs/	n.	鼠标	(Unit 1)

243

multimedia /ˈmʌltiˈmiːdiə/	n.	多媒体；多媒体的采用	(Unit 12)
multiply /ˈmʌltɪplaɪ/	v.	[数] 乘	(Unit 6)
muscle /ˈmʌsl/	n.	肌肉；臂力	(Unit 13)
MySQL		一种开源关系型数据库	(Unit 7)

N

nationality /ˌnæʃəˈnæləti/	n.	国籍	(Unit 7)
naughty /ˈnɔːti/	adj.	顽皮的，淘气的	(Unit 11)
navigate /ˈnævɪgeɪt/	v.	导航	(Unit 8)
near-sighted /ˌnɪəˈsaɪtɪd/	adj.	近视的	(Unit 13)
netter /ˈnetə/	n.	上网冲浪的人	(Unit 12)
nevertheless /ˌnevəðəˈles/	adv.	然而	(Unit 10)
no wonder		毫不奇怪	(Unit 1)
normalisation /ˌnɔːməlaɪˈzeɪʃn/	n.	规范化	(Unit 7)
normally /ˈnɔːməli/	adv.	正常地；通常地	(Unit 11)
not ... at all		一点也不	(Unit 1)
not only ... but also ...		不仅……而且……	(Unit 8)
numerical /njʊ(ː)ˈmerɪkl/	adj.	数字的，用数表示的	(Unit 6)
numerous /ˈnjuːmərəs/	adj.	许多的，无数的	(Unit 2)

O

on demand		一经请求	(Unit 14)
on the other hand		另一方面	(Unit 10)
on the way		在途中	(Unit 4)
one after another		一个接一个，依次的	(Unit 4)
open source		开放资源	(Unit 7)
operate /ˈɒpəreɪt/	v.	操作，开动；动手术	(Unit 3)
opposite /ˈɒpəzɪt/	n.	对立物，相反的事物	
	adj.	对面的；对立的，相反的	(Unit 3)
optical /ˈɒptɪkl/	adj.	光学的；眼的，视力的	(Unit 2)
organise /ˈɔːgənaɪz/	v.	组织	(Unit 4)

Words & Expressions

outlook /ˈaʊtlʊk/	n.	外观	(Unit 8)
output /ˈaʊtpʊt/	n. & v.	输出	(Unit 2)
overall /ˌəʊvərˈɔːl/	adj.	总体的，全面的	(Unit 8)
overhead /ˌəʊvəˈhed/	adj.	头顶的	(Unit 8)
owe sb a favour		欠某人一个情	(Unit 9)

P

passer-by /ˌpɑːsəˈbaɪ/	n.	(pl. passers-by) 过路人	(Unit 10)
password /ˈpɑːswɜːd/	n.	密码，口令	(Unit 12)
perform /pəˈfɔːm/	v.	运行	(Unit 6)
perform /pəˈfɔːm/	v.	执行；表演	(Unit 3)
personal /ˈpɜːsənl/	adj.	个人的	(Unit 1)
perspective /pəˈspektɪv/	n.	看法，视角	(Unit 14)
pirated /ˈpaɪrɪtɪd/	adj.	盗版的	(Unit 3)
pleasure /ˈpleʒə(r)/	n.	乐事	(Unit 9)
plenty /ˈplenti/	n.	丰富；大量	(Unit 13)
plenty of		很多	(Unit 13)
plug in		插入	(Unit 4)
point to		指向	(Unit 4)
pointer /ˈpɔɪntə(r)/	n.	光标；指针	(Unit 4)
pop-up /ˈpɒp ʌp/	adj.	（计算机窗口）弹出的	(Unit 1)
port /pɔːt/	n.	端口	(Unit 4)
poster /ˈpəʊstə(r)/	n.	海报	(Unit 9)
powerful /ˈpaʊəfl/	adj.	强大的	(Unit 1)
prefer /prɪˈfɜː(r)/	v.	更喜欢；宁愿	(Unit 11)
present /prɪˈzent/	v.	呈现	(Unit 4)
presentation /ˌpreznˈteɪʃn/	n.	介绍；陈述	(Unit 3)
printable /ˈprɪntəbl/	adj.	可打印的	(Unit 5)
privacy /ˈprɪvəsi/	n.	隐私	(Unit 13)
process /ˈprəʊses/	v.	处理	(Unit 1)
processor /ˈprəʊsesə(r)/	n.	处理机，处理器	(Unit 2)

245

productive /prə'dʌktɪv/	adj.	有成效的；生产的，多产的	(Unit 13)
program /'prəʊgræm/	n.	程序；节目	(Unit 7)
projector /prə'dʒektə(r)/	n.	投影仪	(Unit 2)
proper /'prɒpə(r)/	adj.	合适的	(Unit 5)
property /'prɒpəti/	n.	资产，财产	(Unit 13)
protocol /'prəʊtəkɒl/	n.	（数据传输的）协议	(Unit 12)
publication /ˌpʌblɪ'keɪʃn/	n.	出版物	(Unit 9)
publish /'pʌblɪʃ/	v.	出版，刊印；公布	(Unit 3)

R

random /'rændəm/	adj.	随机的，任意的	(Unit 2)
ranking /'ræŋkɪŋ/	n.	顺序；等级	(Unit 6)
read-only /'riːd'əʊnli/	n.	只读	
	adj.	只读的	(Unit 2)
receptionist /rɪ'sepʃənɪst/	n.	接待员	(Unit 7)
refer /rɪ'fɜː(r)/	v.	提到，涉及	(Unit 3)
refer to		指的是，说的是	(Unit 3)
regular /'regjələ(r)/	adj.	经常的，频繁的	(Unit 6)
regularly /'regjələli/	adv.	有规律地	(Unit 13)
relational /rɪ'leɪʃənl/	adj.	关系的	(Unit 7)
relationship /rɪ'leɪʃnʃɪp/	n.	关系	(Unit 13)
remote /rɪ'məʊt/	adj.	遥远的	(Unit 10)
remote-control /rɪ'məʊt kən'trəʊl/	n.	遥控；遥控器	(Unit 14)
replace /rɪ'pleɪs/	v.	替换	(Unit 5)
result in		导致	(Unit 10)
résumé /'rezjʊmeɪ/	n.	简历	(Unit 5)
retrieval /rɪ'triːvl/	n.	数据检索	(Unit 7)
retrieve /rɪ'triːv/	v.	检索；重新获得	(Unit 7)
revolutionary /ˌrevə'luːʃənəri/	adj.	革命的	(Unit 14)
revolutionise /ˌrevə'luːʃənaɪz/	v.	改革；使完全不同	(Unit 10)
Ribbon /'rɪbən/	n.	功能区	(Unit 5)

Words & Expressions

right now		立即	(Unit 14)
right-click	n. & v.	右击；击右键	(Unit 4)
ring /rɪŋ/	n.	环形物；环状	(Unit 11)
router /ˈruːtə(r)/	n.	路由器	(Unit 1)

S

sample /ˈsɑːmpl/	n.	样本	(Unit 8)
satisfy one's needs		满足某人的需求	(Unit 8)
scan /skæn/	v.	扫描	(Unit 10)
scanner /ˈskænə(r)/	n.	扫描器，扫描仪	(Unit 2)
score /skɔː(r)/	n.	得分	(Unit 6)
screen /skriːn/	n.	屏幕	(Unit 1)
scroll bar	n.	滚动条	(Unit 4)
search engine		搜索引擎	(Unit 4)
seasonal /ˈsiːzənl/	adj.	季节的	(Unit 5)
secretary /ˈsekrətri/	n.	秘书	(Unit 1)
security /sɪˈkjʊərəti/	n.	安全	(Unit 13)
select /sɪˈlekt/	v.	选择，挑选	(Unit 4)
senior /ˈsiːniə(r)/	adj.	年长的；高级的	(Unit 5)
sense /sens/	n.	感觉	(Unit 6)
server /ˈsɜːvə(r)/	n.	服务器	(Unit 2)
set up		建立，树立	(Unit 7)
shape /ʃeɪp/	n.	形状	(Unit 1)
share with		与……分享	(Unit 3)
sheet /ʃiːt/	n.	（一）张	(Unit 6)
shortcut /ˈʃɔːtkʌt/	n.	捷径	(Unit 4)
show ... around		带……参观	(Unit 2)
shut ... down		（使机器等）关闭	(Unit 2)
sight /saɪt/	n.	看见	(Unit 9)
sign up for		报名参加	(Unit 12)
Silk Road		丝绸之路	(Unit 8)

247

silver /'sɪlvə(r)/	adj.	银发的	(Unit 12)
silver netter		上年纪的上网者	(Unit 12)
similar /'sɪmələ(r)/	adj.	相似的	(Unit 12)
single-click	n. & v.	单击	(Unit 4)
slide /slaɪd/	n.	幻灯片	(Unit 8)
slide /slaɪd/	v.	滑动	(Unit 1)
smart classroom		智慧教室	(Unit 10)
smartphone /'smɑːtfəʊn/	n.	智能手机	(Unit 1)
snail /sneɪl/	n.	蜗牛	(Unit 12)
snail mail		通过邮局寄的信件	(Unit 12)
so far		到目前为止	(Unit 10)
solve /sɒlv/	v.	解决	(Unit 6)
source code		源代码	(Unit 7)
specifically /spə'sɪfɪkli/	adv.	特意；专门地	(Unit 9)
spreadsheet /'spredʃiːt/	n.	电子制表软件；电子数据表	(Unit 3)
spur /spɜː(r)/	n.	支线	(Unit 11)
SQL		结构化查询语言	(Unit 7)
SQL Server		**SQL** 服务器	(Unit 7)
stand for		代表	(Unit 1)
state /steɪt/	v.	陈述，声明	(Unit 7)
statement /'steɪtmənt/	n.	结算单，报表	(Unit 4)
statistical /stə'tɪstɪkl/	adj.	统计的	(Unit 6)
storage /'stɔːrɪdʒ/	n.	存储；贮藏库	(Unit 2)
stretch /stretʃ/	v.	伸展	(Unit 13)
stylus /'staɪləs/	n.	触控笔	(Unit 1)
subtraction /səb'trækʃn/	n.	减去，消减	(Unit 6)
such as		例如，像	(Unit 1)
sudden /'sʌdn/	n.	突然发生的事(只用于习语)	
	adj.	突然的	(Unit 6)
suffer /'sʌfə(r)/	v.	受苦	(Unit 13)
supercomputer /'suːpəkəmpjuːtə(r)/	n.	超级计算机	(Unit 1)

Words & Expressions

surf /sɜːf/	v.	（互联网）冲浪，浏览	(Unit 1)
surfer /ˈsɜːfə(r)/	n.	冲浪者	(Unit 10)
survey /ˈsɜːveɪ/	n.	民意调查	(Unit 6)

T

tablet /ˈtæblət/	n.	平板电脑	(Unit 1)
take up		占用，占据	(Unit 7)
talk with		与……交谈	(Unit 1)
technical /ˈteknɪkl/	adj.	技术的	(Unit 9)
technician /tekˈnɪʃn/	n.	技术人员	(Unit 9)
technology /tekˈnɒlədʒi/	n.	技术	(Unit 8)
temperature /ˈtemprətʃə(r)/	n.	温度	(Unit 14)
template /ˈtempleɪt/	n.	模板	(Unit 8)
terrific /təˈrɪfɪk/	adj.	(口) 好极了	(Unit 6)
the other way round		从相反方向	(Unit 10)
theft /θeft/	n.	偷，行窃	(Unit 13)
throughout /θruːˈaʊt/	prep.	自始至终	(Unit 9)
time-saving /ˈtaɪm seɪvɪŋ/	adj.	省时的	(Unit 5)
title /ˈtaɪtl/	n.	名称；标题	(Unit 5)
to one's surprise		出乎……的意料	(Unit 11)
too .... to		太……以至于不……	(Unit 9)
toolbar /ˈtuːlbɑː(r)/	n.	工具栏	(Unit 5)
topic /ˈtɒpɪk/	n.	主题	(Unit 8)
topology /təˈpɒlədʒi/	n.	拓扑 (网络布局结构)	(Unit 11)
touch /tʌtʃ/	n. & v.	触摸；联系	(Unit 2)
track /træk/	n.	踪迹	(Unit 2)
trial /ˈtraɪəl/	n.	试用	(Unit 3)
trial period		试用期	(Unit 3)
turn ... into		把……转变成	(Unit 2)
turn on		打开 (电视、计算机等)	(Unit 4)
turn out		结果是，被证明是	(Unit 3)

turn to		转向	(Unit 11)
typographic /ˌtaɪpəˈgræfɪk/	adj.	印刷上的，排字上的	(Unit 9)

U

underline /ˌʌndəˈlaɪn/	v.	在……下面画线	
	n.	下画线	(Unit 5)
unexpected /ˌʌnɪkˈspektɪd/	adj.	意想不到的	(Unit 5)
unit /ˈjuːnɪt/	n.	单元	(Unit 1)
update /ˌʌpˈdeɪt/	v.	更新	(Unit 6)
upgrade /ˌʌpˈgreɪd/	v.	升级	(Unit 1)
upper-right	adj.	右上的	(Unit 4)
USB (Universal Serial Bus)		通用串行总线(一种数据通讯方式)	(Unit 4)
USB flash disk		U盘	(Unit 2)

V

value /ˈvæljuː/	n.	[数]值	(Unit 6)
variety /vəˈraɪəti/	n.	各种	(Unit 11)
via /ˈvaɪə, ˈviːə/	prep.	通过	(Unit 11)
video conferencing		视频会议	(Unit 11)
visual /ˈvɪʒuəl/	adj.	视觉的	(Unit 8)
vivid /ˈvɪvɪd/	adj.	生动的	(Unit 10)
voice recognition		语音识别	(Unit 1)

W

walk/walks of life		行业；职业	(Unit 10)
web page /web peɪdʒ/	n.	网页	(Unit 9)
website /ˈwebsaɪt/	n.	网站	(Unit 3)
well-structured /welˈstrʌktʃəd/	adj.	结构良好的	(Unit 7)
whiteboard /ˈwaɪtbɔːd/	n.	白板	(Unit 8)
Wi-Fi /ˈwaɪ faɪ/	n.	无线局域网；无线网络	(Unit 1)
with ease		轻而易举地	(Unit 4)

with the help of		在……的帮助下	(Unit 8)
wonder /ˈwʌndə(r)/	v.	想知道；感到诧异	(Unit 1)
work out		解决，计算出	(Unit 13)
workplace /ˈwɜːkpleɪs/	n.	工作场所，车间	(Unit 10)
workstation /ˈwɜːksteɪʃn/	n.	工作站	(Unit 2)

郑重声明

高等教育出版社依法对本书享有专有出版权。任何未经许可的复制、销售行为均违反《中华人民共和国著作权法》,其行为人将承担相应的民事责任和行政责任;构成犯罪的,将被依法追究刑事责任。为了维护市场秩序,保护读者的合法权益,避免读者误用盗版书造成不良后果,我社将配合行政执法部门和司法机关对违法犯罪的单位和个人进行严厉打击。社会各界人士如发现上述侵权行为,希望及时举报,本社将奖励举报有功人员。

反盗版举报电话　　（010）58581999　58582371　58582488
反盗版举报传真　　（010）82086060
反盗版举报邮箱　　dd@hep.com.cn
通信地址　　北京市西城区德外大街4号
　　　　　　高等教育出版社法律事务与版权管理部
邮政编码　　100120

网络增值服务使用说明

扫描二维码或访问以下链接,查看、下载本书配套资源。

http://2d.hep.cn/1723836/1